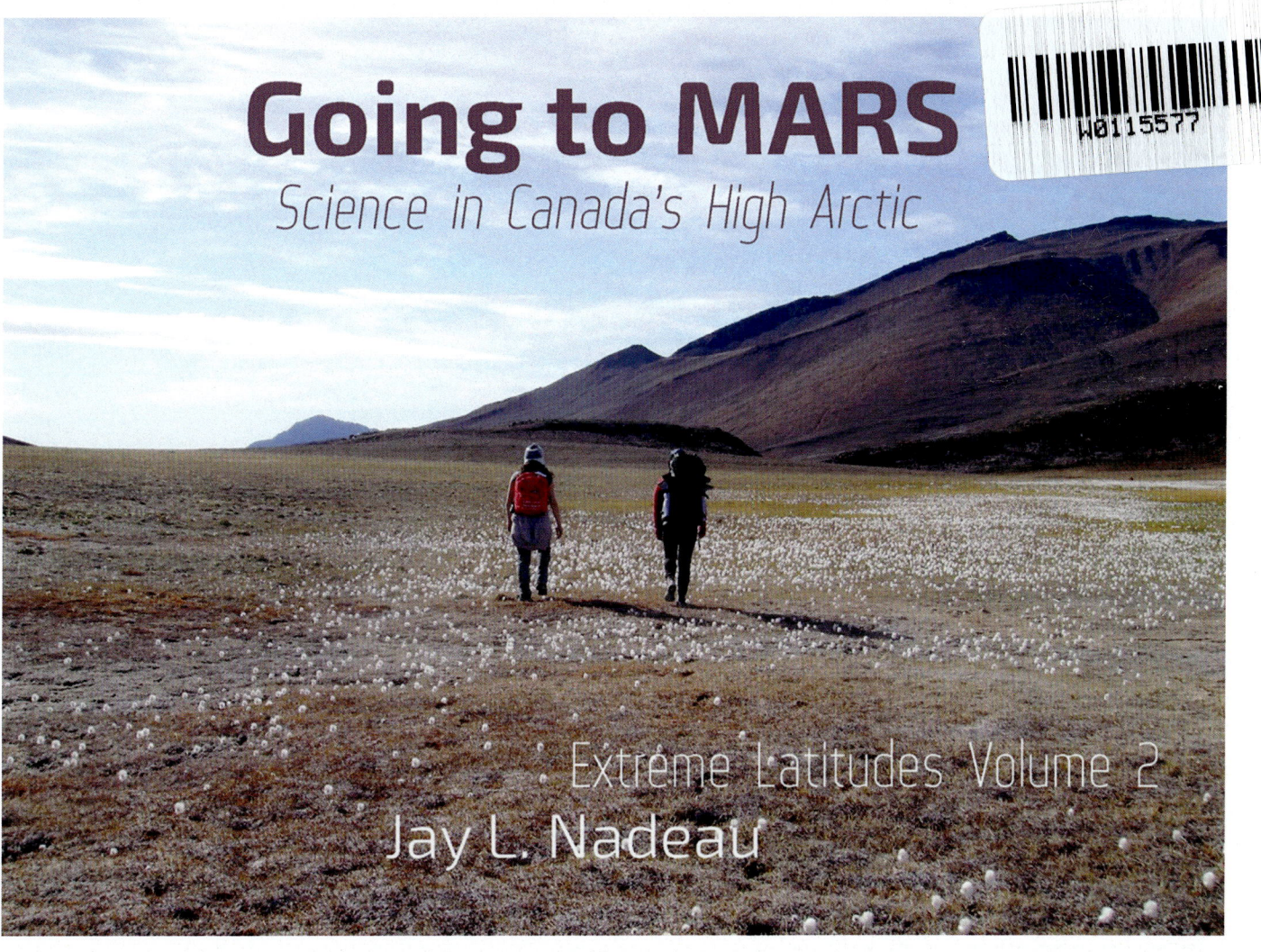

Going to MARS
Science in Canada's High Arctic

Extreme Latitudes Volume 2

Jay L. Nadeau

Published by Bitingduck Press

ISBN 978-1-938463-70-9

eISBN 978-1-938463-71-6

© 2020 Jay Nadeau

All rights reserved

For information contact

Bitingduck Press, LLC

Altadena, CA

notifications@bitingduckpress.com

http://www.bitingduckpress.com

Table of Contents

The photos I in this book were taken over several summer seasons in the years 2006-2009, though the research sections nad references were updated to 2020. The book focuses mainly upon the research of the author and close collaborators. Many other researchers also work on MARS, and many changes have occurred since these trips took place, both in organization and stucture of research and in the featured communities. What is shown here should be considered a historical snapshot of this ever-changing environment.

Chapter 1: People and Places

The McGill Arctic Research Station (MARS) was built in 1960, and is one of the oldest research stations in the High Arctic. It has operated continuously since its construction, providing some of the most complete data on climate, glaciology, and geography in this extreme region. MARS is located on Axel Heiberg Island, Nunavut, Canada, just east of Ellesmere Island and spanning 79-81 degrees north latitude. It is the 31st largest island in the world, and the 7th largest in Canada, covering an area of 43,178 square kilometers (about the size of Switzerland). It has been inhabited by Native people in the past, but has been uninhabited since at least 1900. The nearest permanent community is Grise Fjord on southern Ellesmere Island.

The island is one-third polar desert, one-third glaciated with a total of 1100 glaciers, and one-third mountains (the Princess Margaret range, which rises over 6000 feet high). Although technically desert, its non-glaciated regions host a rich flora and fauna that varies widely from year to year. Ground-nesting birds, arctic hares, foxes, wolves, musk oxen, caribou, and a wide variety of insects populate the tundra. The only woody plant is the arctic willow, restricted to several inches high, but flowers and grasses grow in abundance. The climate is arctic, with average July temperatures near freezing, although Chinook-style winds can lead to highs in the mid-twenties Celsius. The area shows an overall warming trend. Axel Heiberg is also home to an Eocene fossilized forest that suggests that the entire island once enjoyed a warm climate, although with three months of darkness in winter. This extraordinary finding, along with detailed climate records for over a half century, make this island one of the most interesting areas on Earth for climate modeling.

Axel Heiberg was first explored by the Norwegian explorer Otto Sverdrup during the Second Norwegian Polar Expedition (1898-1902). Sverdrup claimed the island for Norway and named it after one of the sponsors of the expedition, Count Axel Heiberg. The three islands he discovered during that expedition—Axel Heiberg, Amund Ringnes, and Ellef Ringnes—are now collectively called the Sverdrup islands. Sverdrup's claims did not go unchallenged. Robert Peary disputed the discovery, and Canada claimed sovereignty over all Arctic islands north of its territory, discovered or undiscovered. Norway finally ceded the Sverdrup Islands to Canada in 1930. Sverdrup sold his expedition records to the Canadian government, but it wasn't until after World War II that Canada began to explore the Arctic in earnest.

In 1955, the Geological Survey of Canada launched "Operation Franklin." From a base at Resolute (see Chapter 3), two geologists in helicopters made two traverses across Axel Heiberg. These studies produced general reports describing the ice cover and geography of the island that sparked the interest of Fritz Müller from McGill University. Müller's first expedition, the Jacobsen-McGill Arctic Research Expeditions, established MARS, and he would continue to visit and study the island until he died in 1980. A glaciologist from Switzerland, he established the quantitative glacier studies that would characterize work on MARS for decades, as well as encouraging research in other disciplines. Although he left Canada to return to Switzerland in 1970, he continued to visit the Canadian Arctic until his sudden death at age 54. The large ice cap in the center of Axel Heiberg is named the Müller ice cap in his honor.

More recently, in the 1990s and later, Axel Heiberg has drawn the attention of geomicrobiologists and astrobiologists as and "analog site"for testing instruments and techniques for discovery of life elsewhere in the Solar System. Most of the rest of our Solar System is cold, so the polar environments are the best models that we have for how life might appear elsewhere. The climate models of Mars suggest that the Red Planet was once much warmer and wetter than it is now, but even this "warm, wet Mars" was cold and dry compared to most sites on Earth. More recently, scientists have become intrigued by the possibility of microbial life on the moons of Jupiter and Saturn. These sites have abundant liquid water under thick ice sheets and prevailing cold temperatures. The acronym "MARS" is just a coincidence, but a fortuitous one!

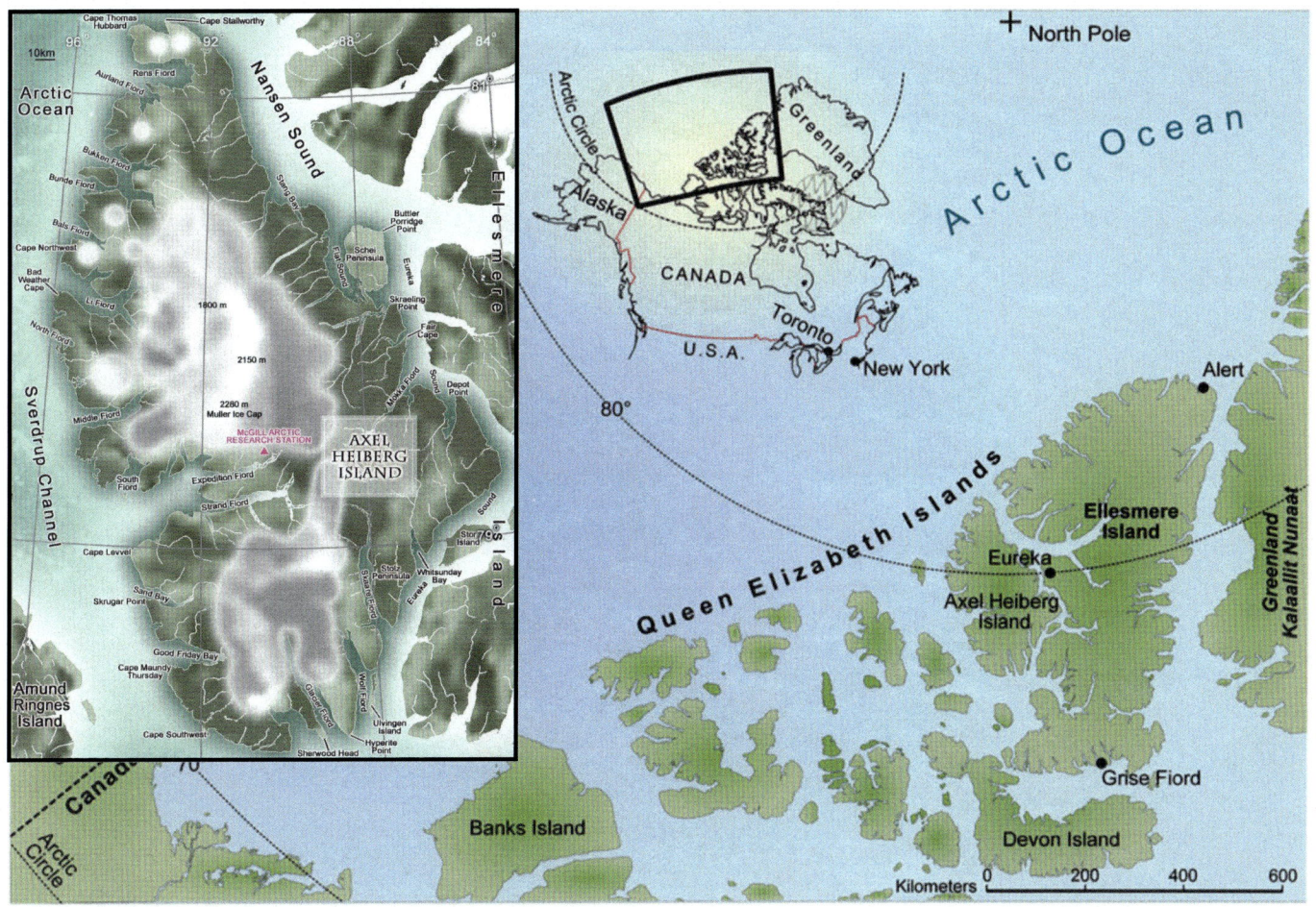

Map of Arctic Canada showing the Queen Elizabeth Islands, which all lie above the Arctic Circle. Ellesmere Island is the largest of these islands. The remainder of the islands are grouped into the Sverdrup and Parry Islands; the Sverdrip Islands include Axel Heiberg and many smaller islands. Devon Island, also used as a Mars analog site, is one of the Parry Islands. The inset shows a detailed map of Axel Heiberg and the location of the McGill Arctic Research Station next to Expedition Fjord.

View over Thompson Glacier, Axel Heiberg Island, Nunavut, Canada July

Devon Island

The Canadian Analogue Research Network (CARN)

Pavilion Lake

M.A.R.S.

The Canadian Analogue Research Network (CARN) was funded by the Canadian Space Agency from 2006-2010 to provide logistical support at selected extraterrestrial analogue sites; the three sites chosen are shown in the figure. The goal was to support understanding of the Solar System and development of technologies and approaches for safe exploration. (Devon Island: photo credit, NASA/JPL/ASU; Pavilion Lake: photo credit NASA; map of Canada: modified from Anchjo, Wikimedia Commons).

ENCELADUS

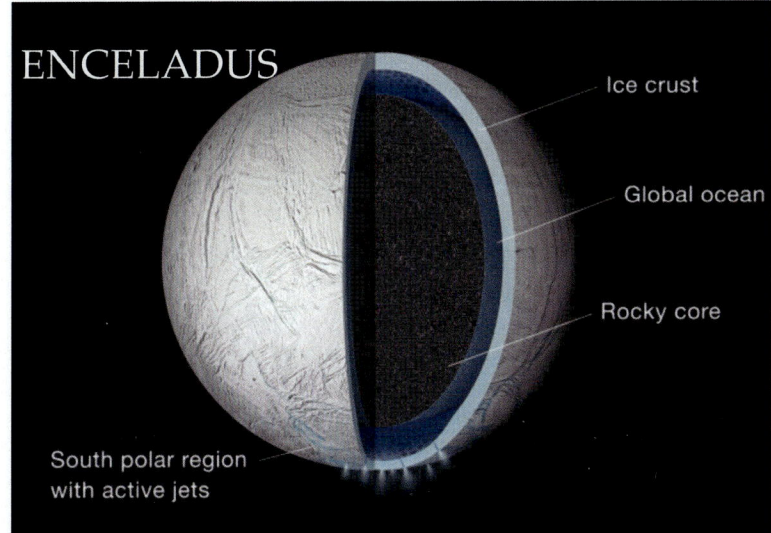

Ice crust

Global ocean

Rocky core

South polar region
with active jets

Areas of the Solar System considered to be he most likely habitats for extant microbial life. Enceladus is a moon of Saturn, Europa is a moon of Jupiter, and Mars contains ice-rich "special regions" that have not yet been explored. (Image credits: NASA/JPL).

EUROPA
with artist's concept of flyby mission

Artist's concept of habitats for life on Enceladus

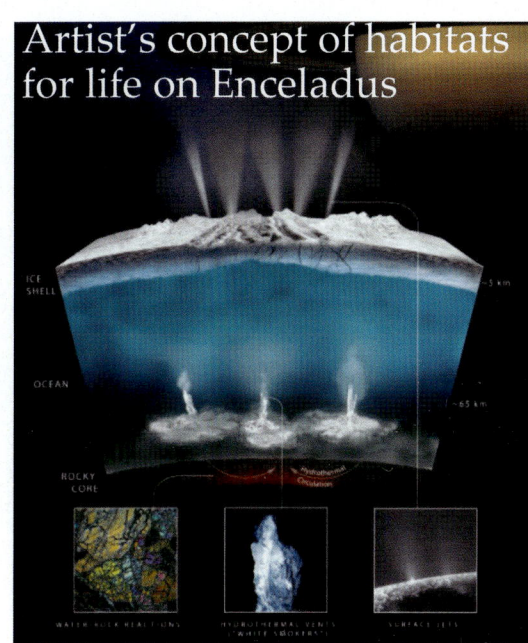

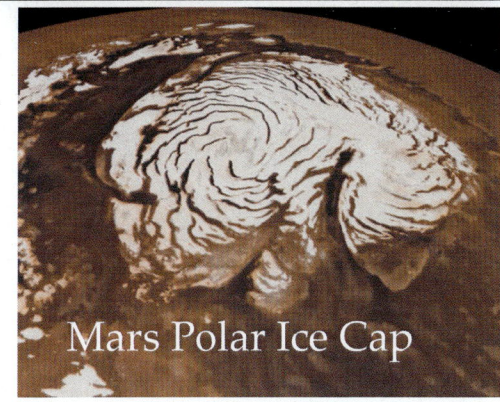

Mars Polar Ice Cap

View from MARS, July

View from MARS, May

Despite having extremely cold temperatures in common, the Arctic and Antarctic are very different. Individual islands in the same Arctic archipelago are also very different. Some areas are desert; some are simultaneously desert and wetland; others are pure ice cap. Seasonal melt produces numerous environments for microbial life that may or may not exit the following year. Bodies of water that never freeze and that support year-round microbial growth are also present. Because visiting MARS in the depths of winter (September through March) is not possible, we can only guess what happens to many of these sites during the cold, dark months.

A seasonal pool on Axel Heiberg Island in July, showing multi-colored mats rich with microbial life

A fluorescence microscopic image of algae from the pool. Each cell is approximately one-one-hundred-thousandth of a meter (10 micrometers) long. The fluorescence comes from dye labeling with a dye specifically designed to target fungal cells, called FUN-1.

Lyle Whyte (left) and his research group study the microbiology of Axel Heiberg (see Chapter 7). This image shows them hard at work in the laboratory that they set up in the MARS primary dwelling structure.

Jay Nadeau's group is developing instruments and techniques for microscopy in extreme field sites and hopefully for exploration of other planets (Chapter 8).

The scenery over the Queen Elizabeth Islands is dramatic and ever-varying. Early July may show pack ice from Resolute on up, or the ice may have retreated completely past the lower half of Axel Heiberg.

The weather at high latitudes is fickle and can change in minutes, even in summer. A busy day of moving equipment around Colour Lake turns into Christmas in July as a snowstorm descends on MARS.

15

Hiking to Thompson Glacier from MARS.

Axel Heiberg Island is home to some of the world's best-studied glaciers. White Glacier (above and below right) and Thompson Glacier (above left and previous page), in the Expedition Fjord area, have been monitored for over half a century. During the 24-long days of summer, meltwater streams down the glaciers, carrying silt, rock, and other debris. The sound is similar to that of a busy freeway as large rocks tumble over each other. Many of these large rocks are so battered by their journey that they will crumble into sand at the slightest touch. Wayne Pollard, shown enjoying a hot drink in the MARS kitchen, has directed the MARS station for decades as well as conducting research into the geophysics of Axel Heiberg.

Following pages: panoramic views of White Glacier (page 18) and Thompson Glacier (page 19).

Aerial view of Axel Heiberg Island in early July.

Aerial view of Axel Heiberg Island in late April.

Expedition Fjord in May

Approaching White Glacier in July

Chapter 2: Extreme Environments

Sea ice breakup in the summer leads to a variety of dramatic patterns. Younger ice looks thinner and more porous than ice that is several years old. Cracks, fissures, and pools appear as the ice melts.

Aerial views of ice on Expedition Fjord

Iceberg glacier is one of very few tidewater glaciers on Axel Heiberg Island, and is the source of numerous icebergs which can be seen floating in open water during the summer.

Steep red mountains, huge glaciers, and deep valleys make the terrain from the air look fully Martian. The rocky hills in front of the glacier are built up from debris that collects as the glacier advances. This terrain, with structures up to hundreds of feet high, is called a "push moraine." The push moraine formed by Thompson glacier has been studied for decades, with a model being developed that describes its formation. It is dotted with ponds in summer that cannot sink into the permafrost below.

Under the midnight sun, the shadows lengthen to create dramatic contrast. This view of the Thompson glacier from the MARS camp at midnight reveals the rugged terrain.

The scale of the features in the push moraine can be appreciated when crossing them on foot. This is the approach to the Thompson glacier on a July day, full of rivulets and rapidly moving streams.

Chilling out on the tundra. The grass grows green and lush in the soil above the push moraine.

At the edge of the glacier in July (upper and lower left) and in May (lower right)

Basking in the sun on an unusually warm July afternoon.

Facing page: MARS in May.

The lake by the MARS station is called Colour Lake. Detailed studies on the ice cover of Colour Lake have been performed for over half a century, recording residual ice and numbers of ice-free summers as well as ice depth. Year-round data have been available since 1992.

A panorama over Colour Lake on a particularly still July day.

July temperatures on Axel Heiberg Island range from below freezing, including snow, to over 20 degrees Celsius (68 degrees Fahrenheit). Katabatic (downhill) winds off the glaciers can lead to rapid rises in temperature. Called foehn winds in the Arctic, these winds are similar to other katabatic winds such as the Chinook (Alberta) or Santa Ana (California), that can lead to increases of 30 degrees C in a single day.

Snow over the MARS airplane landing strip on a July day.

An iceberg in Expedition Fjord.

Next 2 pages: Meltwater pools on glacial ice, called "cryoconite holes," are teeming with microbial life and make intriguing models for possible ecosystems on the icy moons of Jupiter and Saturn.

Chapter 3: Flora and Fauna

The geological features of Axel Heiberg are a striking mix of igneous (volcanic) rock and sedimentary deposits. The volcanic rock is basalt from the Albian age (late Early Cretaceous period, about 100 million years ago) and is known as the Strand Fiord Formation. These rocks are relatively resistant to erosion and show distinct cliffs made of columnar basalt. These are some of the northernmost volcanic rock outcrops on Earth; similar features exist in Scandinavia (Svalbard) and Russia (Franz Josephs Land) and are probably related. The steep cliffs and outcrops of the island are essentially bare of vegetation and are covered by *regolith*: loose, uncompacted dirt, dust and rocks sitting on top of bedrock. Mars and the Moon also have regolith, but only Earth is known to have *soil*, which is the fraction of the regolith that contains organic material and supports plant life.

An igneous formation overlooking Expedition Fjord.

Sedimentary rock formations.

Previous page: a collection of moss and lichens photographed in Resolute.

Axel Heiberg has approximately 137 species of vascular plants and just about as many species of mosses and liverworts. Areas around bodies of water that are sheltered from wind and the most extreme temperatures have the most flora.

Purple saxifrage, *Saxifraga oppositifolia*, is the official flower of Nunavut. It is one of the first plants to bloom in the spring and is an important source of food for Arctic hare.

Saxifrage exudes calcite as a waste product, which is visible as white areas at the tips of the leaves, usually most noticeable in warmer and drier weather.

Mountain avens. Many Arctic plants grow as clumps or cushions, with the current year's growth hard to distinguish from

Dwarf fireweed (*Chamerion latifolium*) is also called "river beauty"since it tends to proliferate around streams, seashores, and flood plains. It is the national flower of Greenland. All parts of the plant are edible and may be eaten as a salad, like spinach, or brewed into a tea.

Alpine arnica (*Arnica alpina*)

A moth and an unusual spider.

The Arctic willow (*Salix arctica*) is Axel Heiberg's only woody plant. It is actually a tree, the northernmost in the world and reduced to less than 10 centimeters high in Axel Heiberg's harsh climate. It serves as a key source of food for hare, muskoxen, and caribou. Willows can be sensitive indicators of climate change, since changing conditions can permit them to grow dramatically taller. Increased summer temperatures lengthen the growing season,and increased winter snowfall can provide more water (important in these polar deserts) as well as protect new growth from blowing sand and snow. The International Tundra Experiment (ITEX) is an international network of researchers examining the impacts of warming on Arctic plants and ecosystems.

Arctic poppy, *Papaver dahlianum* subsp. *polare*.

Arctic cotton.

Muskoxen gather in a circle when threatened, with the large males on the outside protecting the younger and smaller members of the herd in the center. In this case it was the helicopter that caused them to adopt this defensive position.

The hare population varies widely from one year to the next. Sometimes large herds are seen on Axel Heiberg; no one knows the reason that the usually solitary animals decide to congregate.

Eight species of terrestrial mammals inhabit the island: Arctic hare (*Lepus arcticus*), lemming (*Dicrostonyx groenlandicus*), Arctic wolf (*Canus lupus arcticus*), Arctic fox (*Alopex lagopus*), polar bear (*Thalarctos maritimus*), ermine (*Mustela erminea*), muskox (*Ovibos moschatus*), and the endangered Peary caribou. The risk of polar bear encounters is very slight and visitors to the island do not arm themselves. The most any of us has ever seen is a suspected paw print.

An ermine in summer coat.

Axel Heiberg wolves can be elusive. Many times this is the only sign of their presence.

A wolf, photographed from over a mile away with a telephoto lens.

There are roughly 23 species of birds that visit Axel Heiberg Island. including snow goose (Chen caerulescens), gyrfalcon (Falco rusticolus), rock ptarmigan (Lagospus mutus), turnstones (Arenaria interpres), jaegers (Stercorarius longicadus, S. parisiticus), Arctic tern (Sterna paradisaea), snowy owl (Nyctea scandiaca), snow bunting (Plectrophenax nivalis), and three species of gull.

Ruddy turnstones nest on Axel Heiberg, then migrate south along the coasts to winter on the coastlines of six continents. They get their name for their habit of inserting their beaks under stones, shells, or debris to look for food; multiple birds will team up to flip large objects. They are omnivorous, eating insects, spiders, seeds, berries and moss in the Arctic and crustaceans along the shore.

Ruddy turnstone chicks are well camouflaged in the rocks. The parents will feign a broken wing to lure people or other animals away.

The playful Arctic hares love to have their pictures taken and will bound and leap for the camera. They can be a bane to the scientists because they chew electrical cords, ropes, and anything else long and thin.

Chapter 4: Getting There

Doing research on MARS begins with a research proposal to the Canadian Space Agency or to the National Aeronautics and Space Administration and to the Polar Continental Shelf Program, which organizes logistics. Once the project is accepted, travel dates into the field are arranged with several days' leeway on each end. The first stage of the trip takes the groups from Ottawa, Ontario to Iqaluit, the capital of Nunavut. There is usually only a layover of a few hours in Iqaluit, though weather can sometimes delay flights for many days.

The airport is owned by the Government of Nunavut and operated by Nunavut Airport Services. The classic yellow terminal was replaced in 2017.

Iqaluit is the last chance to fix broken equipment or buy other necessities such as toothpaste and batteries.

(Top) Iqaluit panorama showing Frobisher Bay on the right. Iqaluit became the capital of Nunavut in 1999 and has shown remarkable growth since then, with nearly 7800 inhabitants in the 2016 census. It can be reached only by air or water, not by road. Despite its relatively low latitude (below the Arctic Circle), Iqaluit is above the tree line. (Bottom) Aerial views near Iqaluit.

From there, a turboprop or small jet specially equipped for gravel landing delivers the scientists to Resolute Bay, the farthest north inhabited community in Canada. Resolute is a small but fully-operational town, with a school, cooperative store, and several hotels. Because most trips are in July, researchers often get the chance to celebrate Canada Day or Nunavut Day (July 9) in Resolute. The Polar Continental Shelf Program houses the sponsored researchers in a dormitory-style lodging a few kilometers from town; facilities include showers, a wide-screen TV, and laboratory facilities for packing and unpacking of samples, testing instruments, and even rudimentary molecular biology.

The Polar Continental Shelf Program dormitory in Resolute.

Resolute laboratory facilities.

Resolute hangar containing gear of people coming, going, and passing through.

The amount of sea ice still present in Resolute at the beginning of July varies widely from year to year. In a warm year (left), large areas of open water remain. In a more typical year (right), there is still enough ice to be able to walk far out to sea, with only the occasional episode of "floe-hopping."

A sea urchin in the ice

A puppy enoys some cuddle time.

Resolute.

Hiking into the hamlet of Resolute from the research station.

Within the laboratory/hangar building are the logistics planning crew, who give the scientists estimated windows of departure for their flights into the field. These flights are conducted on small Twin Otter planes with large, soft "tundra tires" for landing directly on the terrain on Axel Heiberg. Weather permitting, crews usually spend only 1-2 days in Resolute before departure, but Arctic weather is fickle and the planes operate by visual rules only, so delays are common. The teams have to be ready for their departure to be announced. In just a few minutes, they will have to pack up their gear, leave the cozy Resolute dormitory, and head into the field.

Ready to go from Resolute.

Aerial views of Cornwallis island, departing from Resolute towards the north.

Don't get too comfy!

Sometimes the Twin Otter can land right at MARS, but other times the runway is too soft. In that case, the plane may land miles from the station, and a helicopter carries the gear to MARS while the researchers walk. Either way, once at the station, everyone must gather up their gear and take it to the appropriate location for sleeping, eating, or doing research.

The original MARS cabin sleeps 4-5 people, with most of its space dedicated to a laboratory bench, radio communications, and a stove for heating the air as well as boiling experimental samples. Small, low-power versions of most laboratory instruments have been carefully installed over the years. Up to 20 people stay at MARS at once, with most people sleeping in tents. The brightly colored tent up on the hill was the women's dormitory for this particular season. When sleeping in a tent, there is always light from the midnight sun, bright enough to read by. Some people love this, some people hate it, and some love it for a few days until the sleeplessness starts to catch up with them.

Any food the scientists have brought can be taken to the white kitchen building or the weatherhaven "pantry" on the left. The station is usually well stocked with food and no one needs to bring anything, but that doesn't stop people from having their own secret stashes of calorific items like nut butters and chocolate. The long days of work and many miles covered on foot every day make everyone hungry. Resolute is a dry town, and alcohol at MARS is discouraged.

The kitchen at MARS is well stocked and is left unlocked all year in case a hungry explorer should stumble upon it. It would take a lot for someone to try one of these 60-year-old cans of luncheon meat, which are sometimes the subject of bets or dares.

View of MARS from the hills above

Colour Lake serves as the source of drinking and washing water for the McGill Arctic Research Station. Its low pH (~3.7, about the same as Diet Coke) ensures that few bacteria grow in the water, making it safe to drink. Drinking water is collected in a large plastic garbage bin carried down to the lake by a volunteer each morning. Some summers, but not all, see complete disappearance of the ice cover on the lake. This photo shows an exceptionally warm July. The island's high temperature record was set July 14, 2009.

TO MARS

Colour Lake ice cover during a typical summer day in eary July. Enough open water exists for drinking water collection, but a large amount of ice remains. The dry land bridge serves as a foot path between the runway and MARS.

MARS gear storage weatherhaven.

The sturdy orange-and-white tents are called Weatherhavens (a brand name). They are used at MARS to store gear and food, house people, and set up temporary laboratories.

Surprisingly enough, perishable food products are stored at MARS in a freezer. This is because the extremely intelligent Arctic wolves learn to break into any other food storage receptacles, such as holes in the permafrost. Fresh milk, fruits, and vegetables are delivered by Twin Otter as needed and as weather permits and are stored at ambient temperature in the walk-in tent.

The midnight sun appears high over the mountains, reaching its lowest point at midnight between the bumps of the mountain and then rising again. It is always light enough to work, read, and walk around outside, and there is no chance of seeing the Northern Lights at MARS during the summer—or even the full moon, which appears opposite the midnight sun in the sky and so is always below the horizon.

There are no showers at MARS. Most people simply do without for the seven days to six weeks that they are at the station. Some take pride in returning to civilization looking like yetis. Others volunteer to do dishes for the chance to at least get their hands and nails clean in warm, soapy water. Just about everyone has taken a swim in Colour Lake, but that's usually more for showing off than hygiene.

Sometimes it all becomes too much, and you just need to wash your hair. In that case the meltwater in the moraine, or coming straight off the glacier, will do the trick.

Trekking down to the glacial melt with a basin and a bottle of shampoo. To avoid contaminating the site, all the suds are collected in a basin and disposed of back in camp.

Glacier shower. When sticking your head underneath glacial melt, it's a good idea to have a buddy nearby who can shout "ROCK!" if one happens to tumble down the glacier.

Hauling gear to the "airport."

Breakfast and lunch at MARS are grab and go, but making dinner is a communal task, with different people volunteering each evening to prepare some or all of the meal. Sometimes local fish or game feature in the cuisine, but other times cooks re-create their own comfort food using the ingredients on hand—anything from brats to fish tacos "Nunavut style."

The variety of fresh fruits and vegetables available is at the whim and capacity of the Twin Otters that deliver food and gear. Sometimes a dozen cabbages show up suddenly and need to be used in salad, soup, and casseroles. This red cabbage salad will keep scurvy at bay!

People and animals share scarce resources in the Arctic. The cables to the generators that power life and research on MARS can be very attractive to Arctic hares, which will chew anything long and thin. It is important for researchers at MARS to protect the local wildlife from electrical cables and other technological hazards. It is also important to dispose of food waste and human waste properly. Solid waste (feces) is collected in a garbage bag placed under a toilet seat overlooking the best view in camp. When full, the bags are burned along with other garbage. Urination is performed outside but should be restricted to specific locations, since it upsets the nitrogen balance of the tundra and affects the vegetation. Grass grows lush and green in the "pee spots" and in areas where an animal has died, since both urine and bodies provide a rich source of scarce nitrogen.

Using a fat-tired bike with a trailer to haul gear from MARS to the field site.

A hare poses in front of the mountains.

The red flag provides some privacy for the only outhouse at the station.

The grass is always greener over the dead musk ox.

Baking from scratch using a surprise delivery of sour green apples.

A stroll across a field of Arctic cotton just past Colour Lake.

The "Old Man on the Tundra" is a rock formation is carved with several names, but it's unclear whose. They might be remnants of the original Sverdrup expedition, or even signs of the Hans Kruger team, which disappeared from the area in 1930 and whose final cache was discovered in 2003 on Axel Heiberg. Or they could just be modern graffiti from students or travelers passing by. Either way, it is a strange feeling to find graffiti on an unihabited island, and makes one dwell on the lives and deaths of earlier visitors (see Chapter 9 for some of the mysteries surrounding Axel Heiberg Island).

Chapter 6: Doing Research

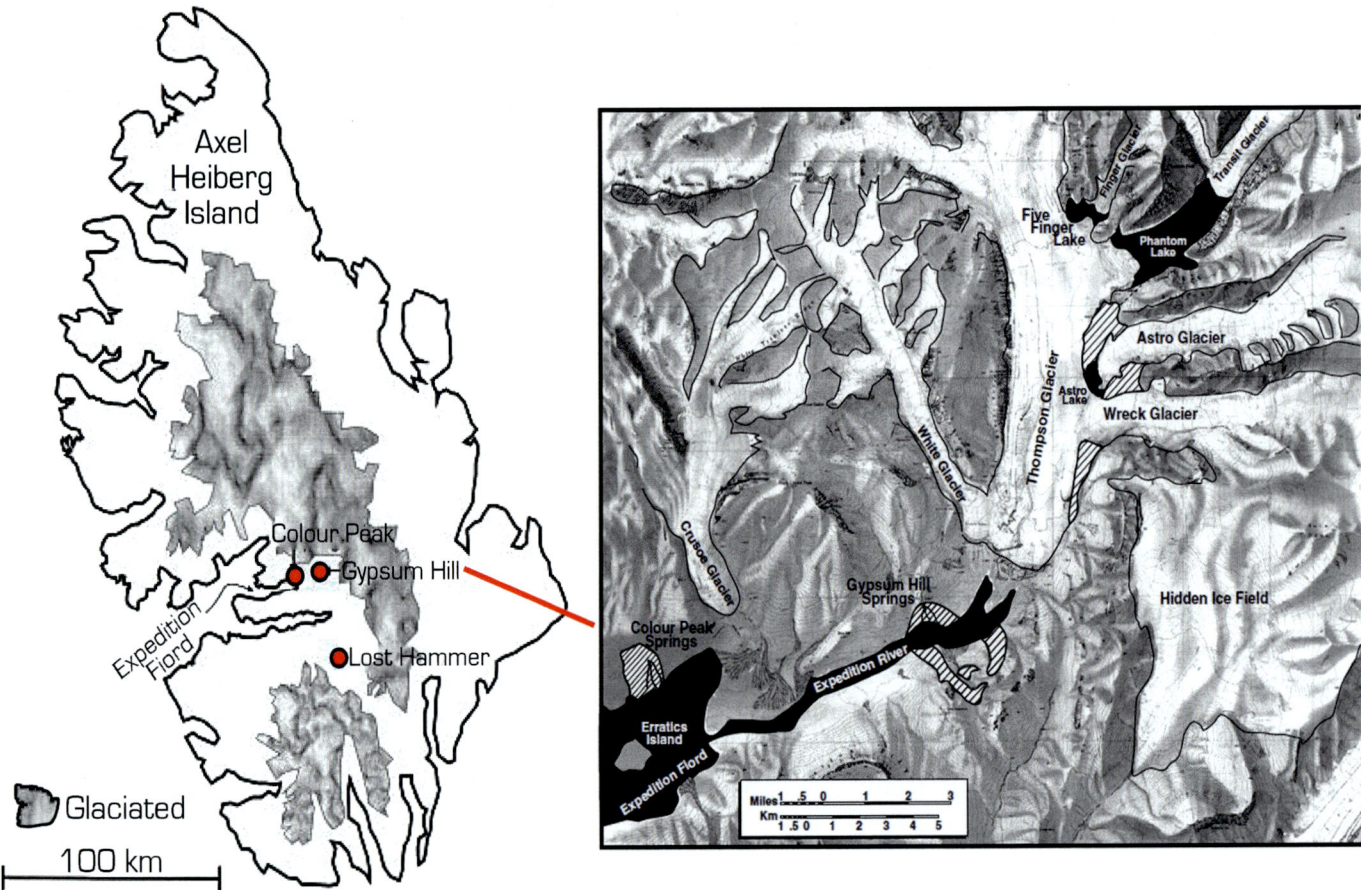

Large- and small-scale maps of the research areas featured in this book. There are other areas of great interest on the island, but we focus here mostly on those of interest as Mars analogs.

Some of the research on Axel Heiberg is done right at the MARS site, such as measurements of carbon dioxide in the soil that indicate microbial activity. Most of the work, however, is done at one or more remote sites that are of particular astrobiological interest. Gypsum Hill is just over the ridge from MARS and is usually accessed on foot, with several people often needed to carry equipment and sampling tubes. Gypsum Hill is just on the edge of Expedition Fjord. Another site of interest is Colour Peak, approximately 15 km from MARS. It is usually accessed by helicopter since walking requires fording streams that are highly treacherous, especially in warmer years. The Colour Peak site resembles Gypsum Hill mineralogically but contains no large microbial populations. Another site is Lost Hammer, which is an extreme cold spring (temperature -4.8 ºC) with a high salt concentration and which emits methane, carbon dioxide, and hydrogen sulfide. Finally, the groups studying glaciers do their work directly on the ice surfaces of one or more of the many glaciers such as White Glacier and Baby Glacier.

Sampling algae with a slide.

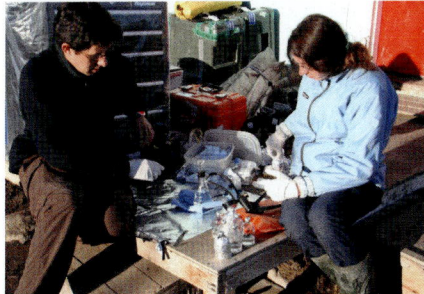
Two students extract DNA from samples of soil and water. They are choosing to work at 2 am because the wind is still, no one is awake to bother them, and of course there's plenty of daylight!

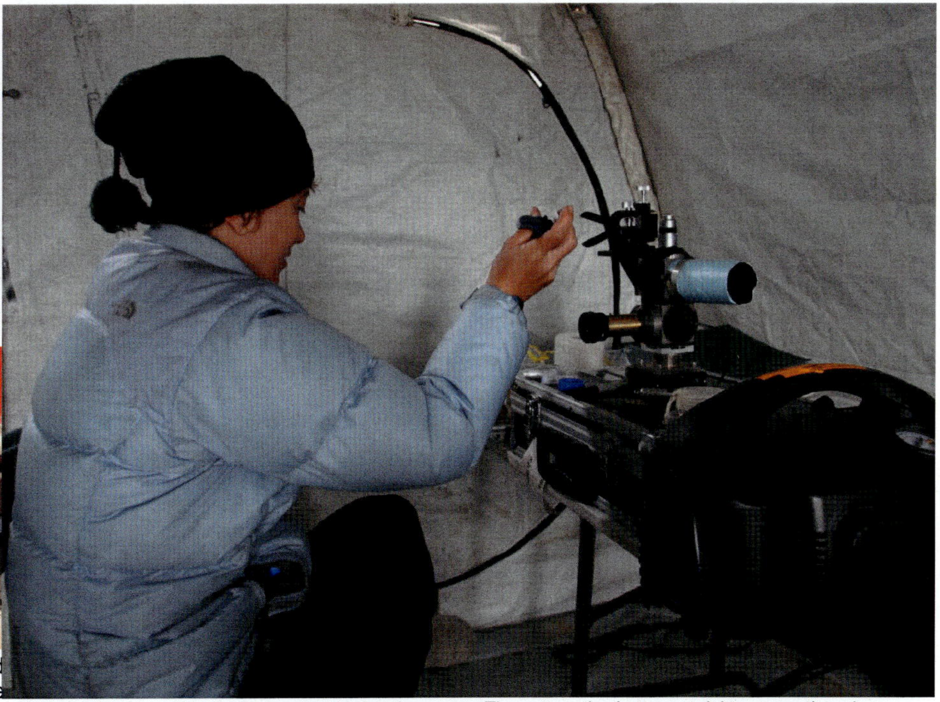

Taking refuge in a Weatherhaven tent to do microscopy. The automotive battery at right powers the microscope light-emitting diode illumination source and camera. Nothing protects the researcher from the cold.

Some sites, such as the Lost Hammer spring shown here, are too difficult to reach on foot because of the extreme terrain on the island. Researchers can schedule helicopter time for hours to days in order to visit these more remote areas.

A helicopter ready to take researchers to their day's destination, with a curious hare checking it out. The large Arctic hares are playful and not afraid of people. In this high Arctic location, they stay white all year round.

Measuring the chemical properties at Lost Hammer.

Gypsum Hill is easily accessed on foot from MARS, and is one of the best studied sites by geochemists, microbiologists, and astrobiologists. There are about 40 springs, first discovered by the Jacobesen-McGill expeditions and described in Müller's original reports.

Gypsum Hill is a collection of unique cold springs that may be as close to Mars as anything on Earth. The water in the springs passes through 600 meters of permafrost, then emerges cold, anoxic (lacking oxygen), and filled with sulfur. Its high concentration of salt and other dissolved minerals keep the water from freezing even at temperatures below 0 ºC/32 ºF. The springs remain liquid even in the depths of winter, when the surrounding air temperature can be below -40 ºC (same as -40 ºF). This makes them a model for how Martian groundwater might survive and harbor life even under current conditions of extreme cold, low atmospheric pressure, and absence of oxygen. The bacteria in the springs thrive on hydrogen sulfide ($H2S$), which gives off a rotten eggs odor that we are evolved to be able to detect in minute quantities because it is highly toxic to all animals. The Gypsum Hill springs are the only known nonvolcanic, hypersaline, sulphidic cold springs on Earth, and among the only springs known that flow through permafrost.

Above and previous page: Gypsum Hill. The white areas here (except for the glacier in the background) are not ice—they are gypsum crystals, hence the name. The yellow areas are elemental sulfur. On this warm summer day, there is no ice or snow at Gypsum Hill, and the surface is soft underfoot.

Each day the destination site and sample and data collection plan are determined. Traveling to the site can take just a few minutes or several hours of foot, ATV, snowmobile, or helicopter travel. Instruments may be carried in suitcases, backpacks, or boxes. The extreme terrain sometimes requires traveling on all fours, so backpacks or at least hiking poles are highly recommended. A mountain bike with very wide, soft tires also works well for traveling over the tundra and quicksand.

Chapter 7: Dominant Lifeforms

Approaching
Thompson glacier
in May

Approaching
Thompson glacier
in July

The Whyte lab has been involved for many years in discovery and microbiological characterization of cold springs on Axel Heiberg Island, and of their use as Mars analogs. One of the earliest investigators of the Gypsum Hill cold springs, Lyle turned his attention to the life in the springs. The main pools of the cold springs are extremely clear, meaning that very few microorganisms live there. But where the water runs out of the pools in shallow channels, bacteria collect into mat- or rope-like formations that can be seen by eye, called biofilms. The biofilms at Gypsum Hill form a lacy pattern that streams in the outflow water, so Lyle's group named them "streamers."

The Whyte group has learned a tremendous amount about the streamers over the past decade. One key aspect of the work is to see how the bacteria that form the streamers survive the harsh winters in the Canadian Arctic, with temperatures down to -50 ºC and even colder. Visiting the island is impossible in the true depths of winter, but the Whyte lab has pushed these limits by visiting in late April to early May. There is already 24-hour daylight at this time, but everything is covered with snow and ice--including Colour Lake, so that drinking water has to be collected with an augur.

Interestingly, what Lyle and his group found was that the streamers grow larger and stronger during the winter than they are in summer. The bacteria are protected by a shell of ice, and the hydrogen sulfide that they respire is concentrated inside the shell.

The first steps in characterizing the streamers were simply to pick them up with tweezers and look at them under the microscope. This revealed a large number of highly active bacteria. Although Lyle's students tried many different methods to grow the bacteria in culture medium or on a Petri dish, they were not successful. The Gypsum Hill streamers are among the 80% of environmental bacteria that are "unculturable," because they require something to grow that microbiologists haven't yet figured out in the laboratory.

Techniques for sequencing or genomics developed rapidly during the years Whyte's group visited the island, so they have performed many analyses on the DNA from the streamers to find out how many organisms of what type live there. They found that the streamers as sampled in May were dominated by *Thiomicrorhabdus*, a genus of bacteria in the phylum Proteobacteria. *Thiomicrorhabdus* makes up 27% of the streamer bacterial community. Other Proteobacteria as well as members of the phyla Actinobacteria, Bacteroidetes, and Cyanobacteria made up the balance.

Most recently, his group is testing methods for sequencing that work in the field. These instruments are a prototype for something that might be used eventually on Mars or Europa.

Although Gypsum Hill may be Lyle's favorite site, it is not his only area of study on Axel Heiberg Island. He has also cultured a variety of samples from around Expedition Fjord in order to identify Arctic microorganisms that are capable of inhibiting the growth of bacteria that cause disease, such as *Staphylococcus aureus*. This uses an innovative culturing device called the cryo-iPlate that permits bulk culturing of a large number of environmental samples.

Lyle is also a pioneer in searching for active microbial life in permanently frozen ice and soil (permafrost). Reports that there were active bacteria in permafrost first came from Russian studies over a century ago. But these studies were controversial because it is very difficult to drill into permafrost without contaminating it with bacteria from the drill, the surrounding soil, the containers used to transport the samples back to the laboratory, or the experimenters' hands and faces. Lyle's group has worked to develop fluorescent tracers that are placed on the drill bit, surrounding soil, and other areas to indicate contamination. This way the contaminated areas of sampled permafrost can be identified and removed, leaving only the pristine core to be examined by microscopy and gene sequencing. These techniques are of great interest for Mars exploration, since the sun's ultraviolet rays destroy all signs of life on the planet's surface. It is believed that any life would have to live several centimeters deep. The drilling techniques developed for permafrost will hopefully one day be carried over to Mars exploration.

Gypsum Hill pool in May

Gypsum Hill pool in July

Gypsum Hill seen across spatial scales.

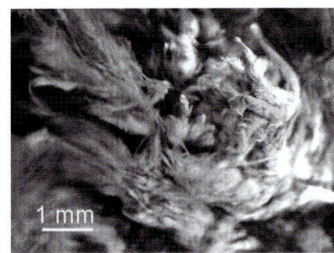

Low-power microscope image of the streamer biofilm. It is flowing in the stream like a lace curtain in a breeze.

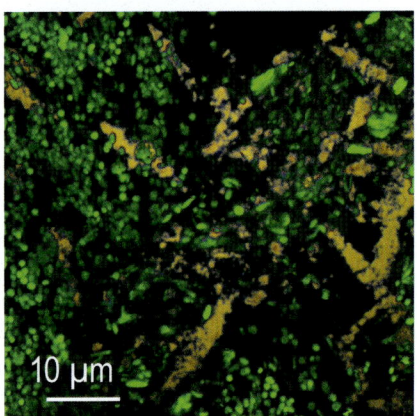

High power fluorescence microscope image. The green indicates bacterial cells and the yellow is sulfur minerals.

The cold spring pools are tens of centimeters across. The outflow channels, where the bacterial biofilm streamers grow, are a few centimeters across. Here Lyle's finger is indicating an example of the streamer biofilm. Would this look like "life" if we saw it on Mars?

Electron microscopic images of elements of the streamer biofilm. For reference, each of the bacterial cells is approximately one micrometer (one one-millionth of a meter) long.

The top image shows the bacteria adhering tightly to a sulfur granule. Note the ridges and clefts in the granule that may be caused by the bacteria. Finding mineral fragments where cells have grown can often show their shape millions of years after the bacteria have died, serving as a biosignature for extinct life.

The tufted minerals at the upper right are also sulfur, but there are no bacterial cells apparent.

The image at the lower right is an unprocessed sample of the streamer biofilm seen by scanning electron microscopy. It is a jumble of different minerals and bacterial cells.

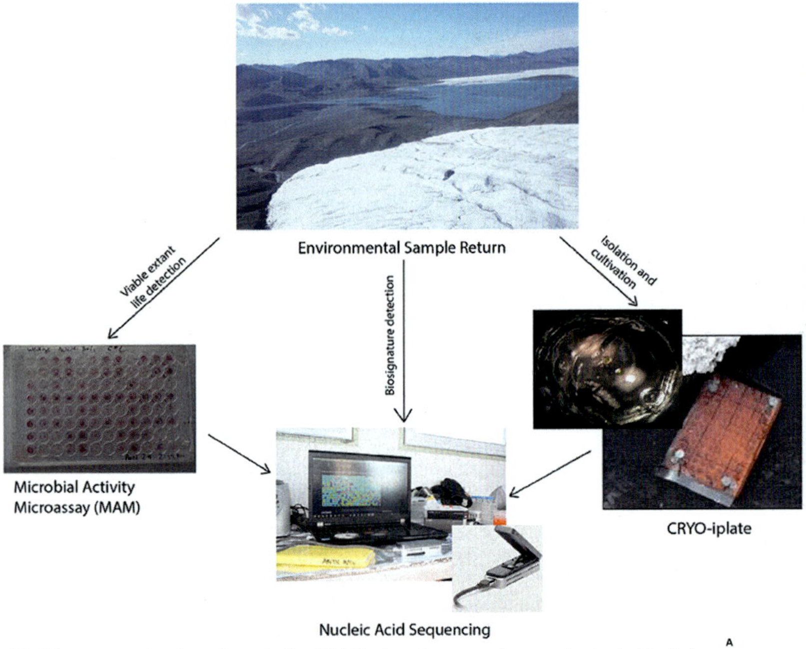

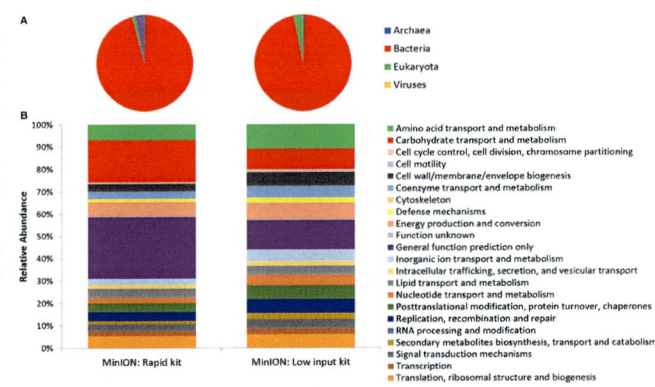
Sampling at Gypsum Hill

Environmental Sample Return

Viable extant life detection

Isolation and cultivation

Biosignature detection

Microbial Activity Microassay (MAM)

Nucleic Acid Sequencing

CRYO-iplate

Sequencing results from the MinION.

Modular system for detection of microbial life in extreme environments, tested by Lyle Whyte's group on MARS. The Microbial Activity Microassay (MAM) (left) produces a colorimetric (purple) signal when microorganisms are present that can metabolize the amino acid L-serine. DNA sequencing using the MinION nanopore system (center) detects nucleic acid biosignatures. The CRYO-iplate (right) is a sterile nutrient medium in multiple small wells. A highly diluted sample, with a dilution chosen to approximate a single bacterial cell, is added to each well and incubated at appropriate temperatures (either directly on the Arctic soil or in the laboratory).

Both data figures are from: Goordial J, Altshuler I, Hindson K, Chan-Yam K, Marcolefas E and Whyte LG (2017) In Situ Field Sequencing and Life Detection in Remote (79°26'N) Canadian High Arctic Permafrost Ice Wedge Microbial Communities. Front. Microbiol. 8:2594. doi: 10.3389/fmicb.2017.02594

A

Archaea
Bacteria
Eukaryota
Viruses

B

Amino acid transport and metabolism
Carbohydrate transport and metabolism
Cell cycle control, cell division, chromosome partitioning
Cell motility
Cell wall/membrane/envelope biogenesis
Coenzyme transport and metabolism
Cytoskeleton
Defense mechanisms
Energy production and conversion
Function unknown
General function prediction only
Inorganic ion transport and metabolism
Intracellular trafficking, secretion, and vesicular transport
Lipid transport and metabolism
Nucleotide transport and metabolism
Posttranslational modification, protein turnover, chaperones
Replication, recombination and repair
RNA processing and modification
Secondary metabolites biosynthesis, transport and catabolism
Signal transduction mechanisms
Transcription
Translation, ribosomal structure and biogenesis

Relative Abundance

100%
90%
80%
70%
60%
50%
40%
30%
20%
10%
0%

MinION: Rapid kit

MinION: Low input kit

Marcolefas E, Leung T, Okshevsky M, McKay G, Hignett E, Hamel J, Aguirre G, Blenner-Hassett O, Boyle B, Lévesque RC, Nguyen D, Gruenheid S and Whyte L (2019) Culture-Dependent Bioprospecting of Bacterial Isolates From the Canadian High Arctic Displaying Antibacterial Activity. Front. Microbiol. 10:1836. doi: 10.3389/fmicb.2019.01836

Samples from many different sediment, soil and rock sites around Expedition Fjord were collected for analysis for anti-microbial activity. ESKAPE refers to bacteria recognized by the Infectious Disease Society of America as posing the most significant risk to public health: *Enterococcus faecium, Staphylococcus aureus, Klebsiella pneumoniae, Acinetobacter baumannii, Pseudomonas aeruginosa,* and *Enterococcus* species.

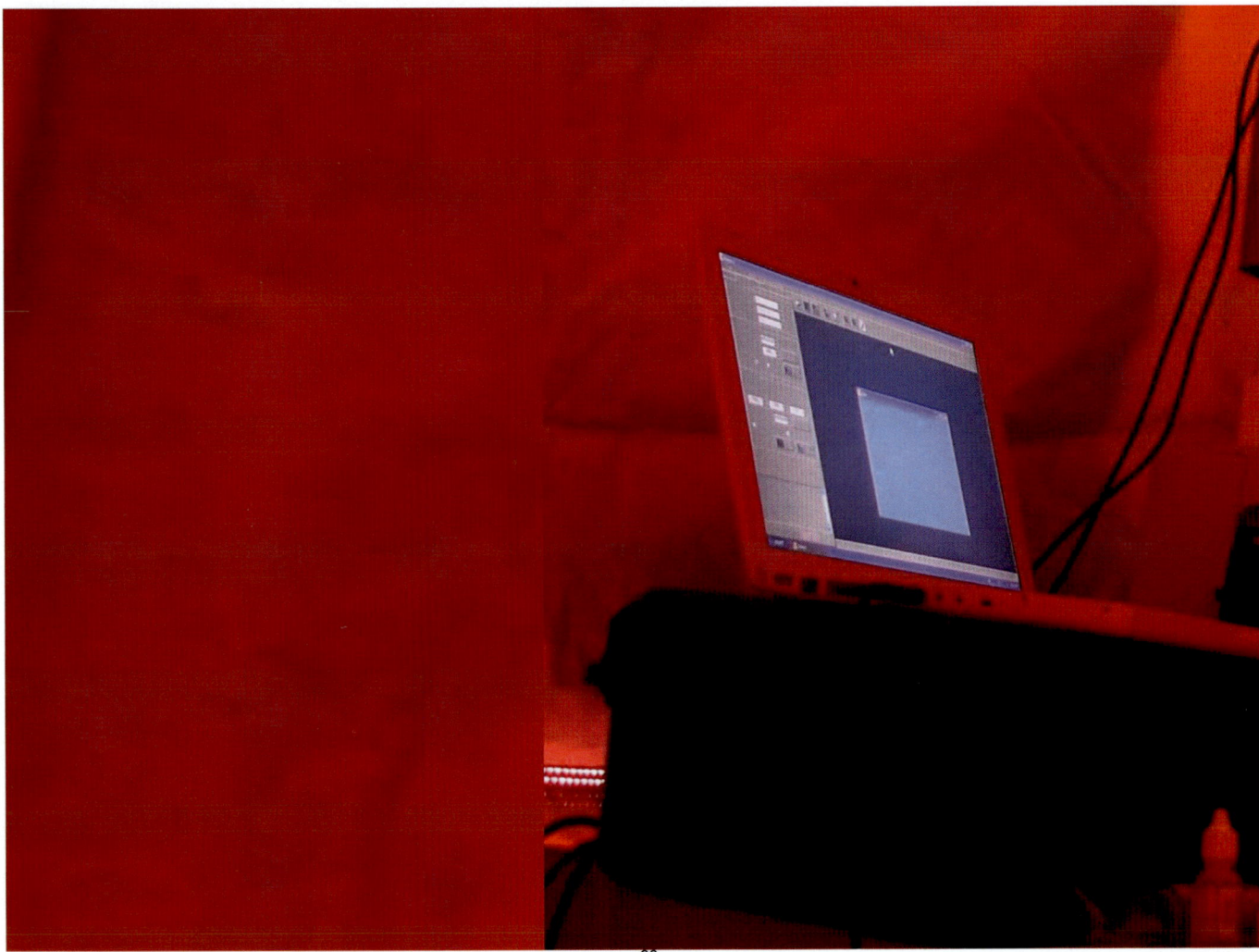

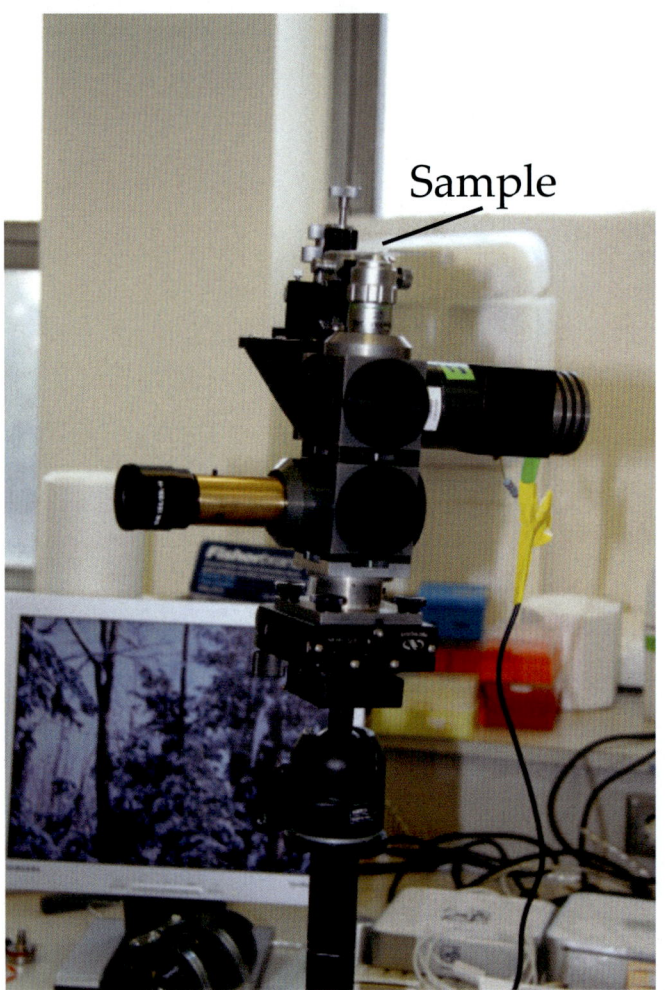

Sample

The Nadeau lab, formerly of McGill University in Montreal and currently at Portland State University in Oregon, is "looking for life by looking for life." Jay's lab builds microscopes and develops microscopy techniques for detection of bacteria in extreme environments, most of them involving ice.

The surest way to practice "life detection" on a sample of Earth soil, rock, water, or ice is to label it with one or more fluorescent dyes that target one or more key biological molecules, and then examine it with high-resolution fluorescence microscopy. These key molecules include (but are not limited to)

- Cell membranes (lipid)
- Glycolipid cell wall components
- Nucleic acids (DNA and RNA).

These molecules are known as "molecular biosignatures" because they are indicators that life exists or has existed.

The use of several dyes—for example, one that labels cell membranes and another that tags DNA—ensures that only cells, and not debris, are detected. These methods are extremely sensitive and have been used to detect low levels of bacterial life in many types of Mars analog samples.

Left: Simple, robust fluorescence microscope designed for field use. The sample sits on a glass slide above a high power objective lens. The eyepiece can be seen on the left. The cubes contain optics for excitation of the dyes using light-emitting diodes (LEDs) of 3 different colors and for filtering the excitation and emission light. All of these elements may be easily swapped out to adjust the colors of excitation and emission needed for a particular experiment.

Following page: Fluorescent biosignatures. (A) A diagram of a bacterial cell showing internal structures that can be targeted with fluorescent dyes. Bacteria do not have a nucleus, but there is a "nucleoid" region containing DNA. (B) Dyes targeting cell walls of bacteria and fungi. (C) Dyes targeting DNA show bacteria on a rock and individual nucleoid regions. (D) Dyes targeting cell walls and capsules of fungi and bacteria. Scales are shown in micrometers (μm).

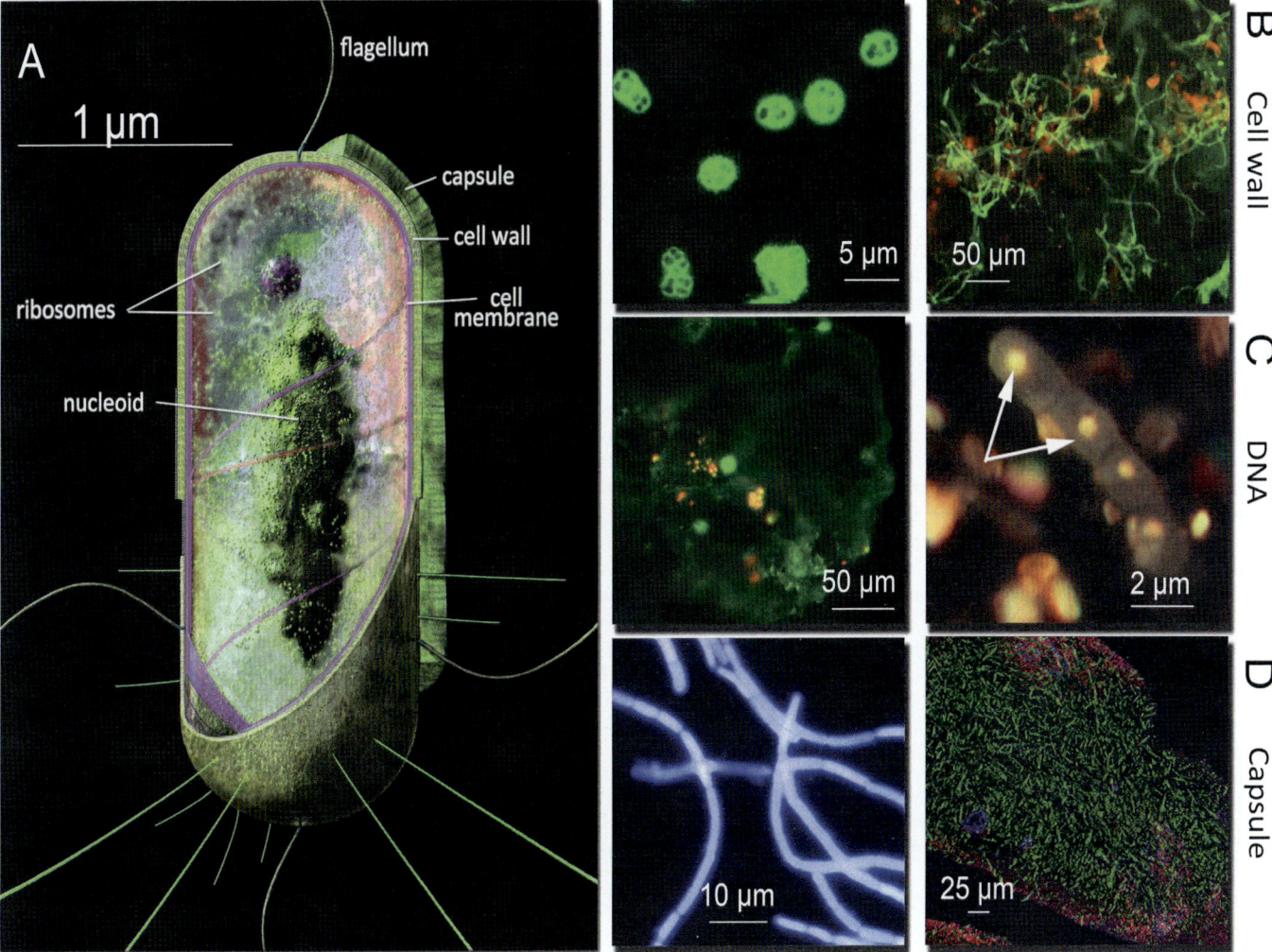

Many astrobiology investigations assume the worst-case life-detection scenario: extraterrestrial life is probably long extinct or sufficiently different from Earth life so that it cannot be detected based upon any Earth chemistry such as lipids, amino acids, or nucleic acids (DNA and RNA). While it is important to develop tools for looking for "strange life," or "life as we don't know it," such an approach isn't necessary on Mars. It is certain that microbial life has traveled between Earth and Mars. This concept is known as "limited panspermia." In the early years of both planets' histories, they were being bombarded by meteors and comets, which carried rocks—probably sheltering bacteria—from on place to the other. Transfer from Mars to Earth is easier than the other way round because of Mars's smaller gravity (1/3 that of Earth).

In this case, Martian microbial life would possess at least some of the same molecular biosignatures as Earth life. Even if some of the molecules were different—for example, something other than DNA used to encode genetic information—other molecules such as lipids would probably be similar enough to work with Earth dyes. Lipids have been called a "universal biosignature" because they have been found in extraterrestrial samples such as comets, where they are not associated with life, but are similar enough to Earth lipids to be detected and dyed. Using microscopy also allows direct inspection to see if objects that look like cells are present. Since all known life consists of cells, it helps answer the question *What does extraterrestrial life look like?*

The problem is that while researchers on Earth use fluorescence microscopy routinely, it is difficult to perform on another planet during a robotic mission. Extraction of bacteria from soil and rock, centrifugation, labeling and washing, and slide preparation are all difficult to do under autonomous robotic control with limited available power and no water except what is brought along. If the bacterial concentration is very low, it is also important to preserve every cell. Separation of particles from a soil sample can often remove most of the microorganisms, especially in silts and clays. The challenges faced by robotic microscopy that aren't seen on Earth include:

- The signal must be unambiguous. No one will dispute bacterial detection in an Earth sample, and no one really cares if there are a million or two million cells per milligram. To claim detection of extraterrestrial life, the signal must be specific and compelling.
- Sophisticated sample preparation steps such as sample washing and centrifugation must be eliminated. An ideal fluorescent tag would have zero background—that is, the probe is nonfluorescent until it binds the molecular biosignature of choice, or it shifts color upon binding. Little nonspecific fluorescent binding to soil or rock particles.
- The microscope itself must be as simple, light, and robust as possible. This means eliminating complex objective lenses and ideally using only a single light source, such as an LED, to excite all of the dyes.

Doing fluorescence microscopy on Axel Heiberg gives just a hint of how hard it will be to do it on Mars. The following page shows some of the challenges and results from the Nadeau lab's trips in 2006 and 2007.

Next page. (A) Sampling the pools at Gypsum Hill. (B) Trying to do field microscopy in a small portable tent to block out sunlight. The tent blew away in the wind. (C) The researchers decided to collect samples instead and take them back to camp. Samples of liquid from Gypsum Hill were collected into sterile vials and placed on ice, then carried back to MARS. (D) At MARS, a Weatherhaven tent was dark enough inside to serve as a micrbiology laboratory. (E) Samples from Gypsum Hill. The green and yellow are DNA-targeting dyes. Most of the small spots labeled by the dyes are bacteria, but the larger, cigar-shaped structures remained ambiguous and constituted a real-life life detection problem. The red cells are algae that fluoresce due to the presence of chlorophyll, no dyes needed.

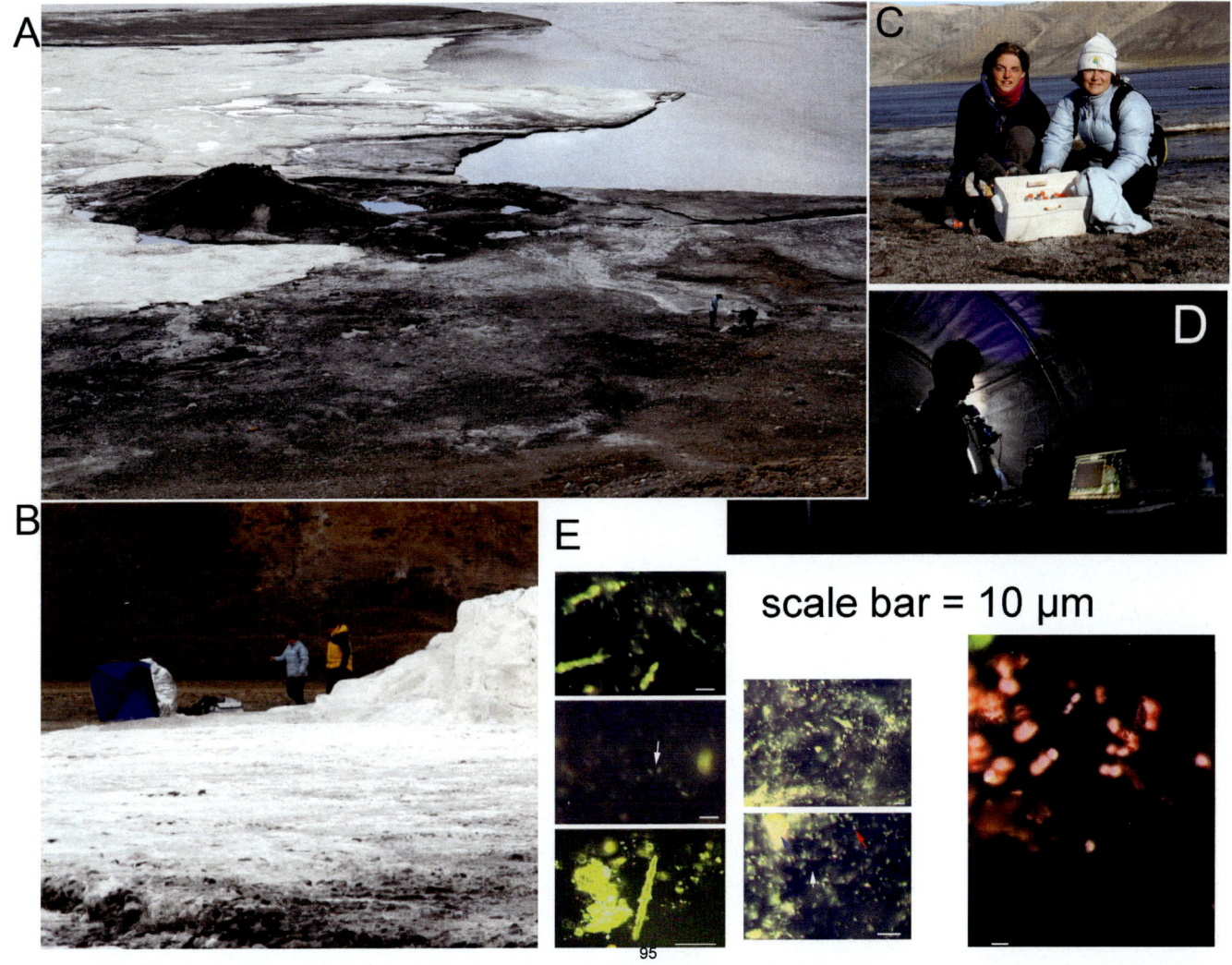

scale bar = 10 μm

95

In 2007 and later, the Nadeau lab began exploring alternative microscopy technologies for automated life detection. While fluorescence microscopy is useful, it requires a good deal of expert operator input. An alternative to traditional microscopy is *holographic microscopy,* which captures a large volume of liquid in a single snapshot and does not require any focusing. The instrument shown here is a holographic microscope encased in an aluminum can specifically designed to float just at the surface of the water.

Pool at Gypsum Hill with holographic microscope. One of the researchers, in the left corner, is shielding his laptop with a jacket to be able to read the screen and obtain real-time information on bacterial swimming.

At Gypsum Hill, the group used the tripod method to avoid stirring up the black sediment at the bottom of the pool. This sediment would get into the microscope and obscure the images.

Deploying the microscope at different depths in different bodies of water gives an idea of the conditions under which microbial life can be detected and when signs of life are missed.

In order to operate completely autonomously, the microscope should be tethered to a robot. In collaboration with Gregory Dudek of McGill and his PhD student Philippe Giguère (pictured; now a professor at Laval University), the Nadeau lab tied the holographic microscope to an amphibious robot that could walk over land and then swim out into bodies of water such as Colour Lake.

Philippe unpacking the robot while Eve Dumas, a Master's student, looks on. Eve is wearing waders so she can stand in the lake in case the robot seems to be getting lost, in which case she'll grab it by hand to rescue it. The robot was Philippe's PhD project and no one was taking any chances with its safety.

Deploying the holographic microscope. The long cables led to a battery to power the laser light source (at one end) and a computer (on the other end). The instrument could be held by hand, suspended from a tripod, or simply dropped in the lake.

Images taken with holographic microscopy

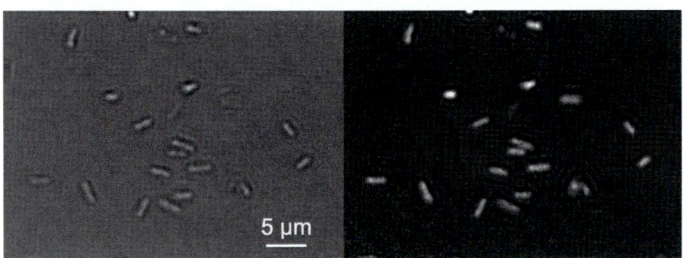

Images of a test strain of bacteria. Holographic imaging results in amplitude images (brighfield, left) and phase images (right). Phase images give the shift in the phase of light as it passes through a specimen, so are a measure of how the specimen's refractive index differs from its surrounding medium (in this case, water).

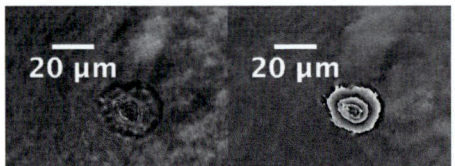

Images of a crystal from the Gypsum Hill cold springs in amplitude (left) and phase (right).

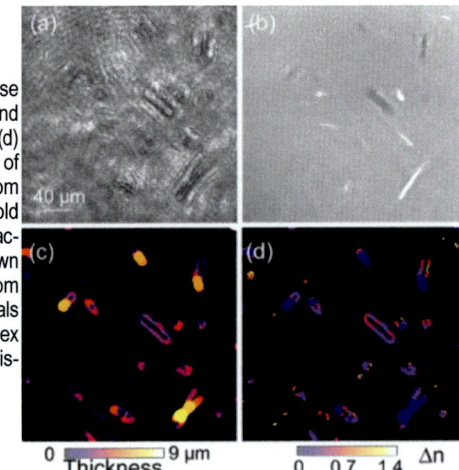

Amplitude (a), phase (b), thickness (c) and refractive index (d) measurements of micro-minerals from the Gypsum Hill cold springs. The refractive index is shown as a difference from water, so the crystals have an actual index of about 2.0, consistent with sulfur.

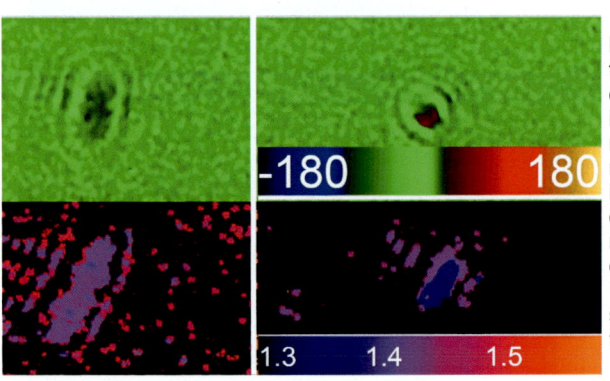

Phase information allows for an estimate of the index of refraction of the sample if the sample thickness is nown. Here a single bacterial cell shows an increase in phase shift as it turns end-on, changing its thickness (the phase shift is indicated in degrees, -180 to 180). The refractive index is shown in shades of purple; the value for water is 1.33.

Tracks of swimming bacteria indicate activity in 3 dimensions.

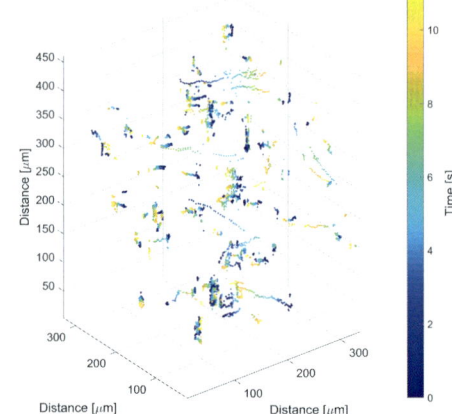

Aerial view of MARS in May

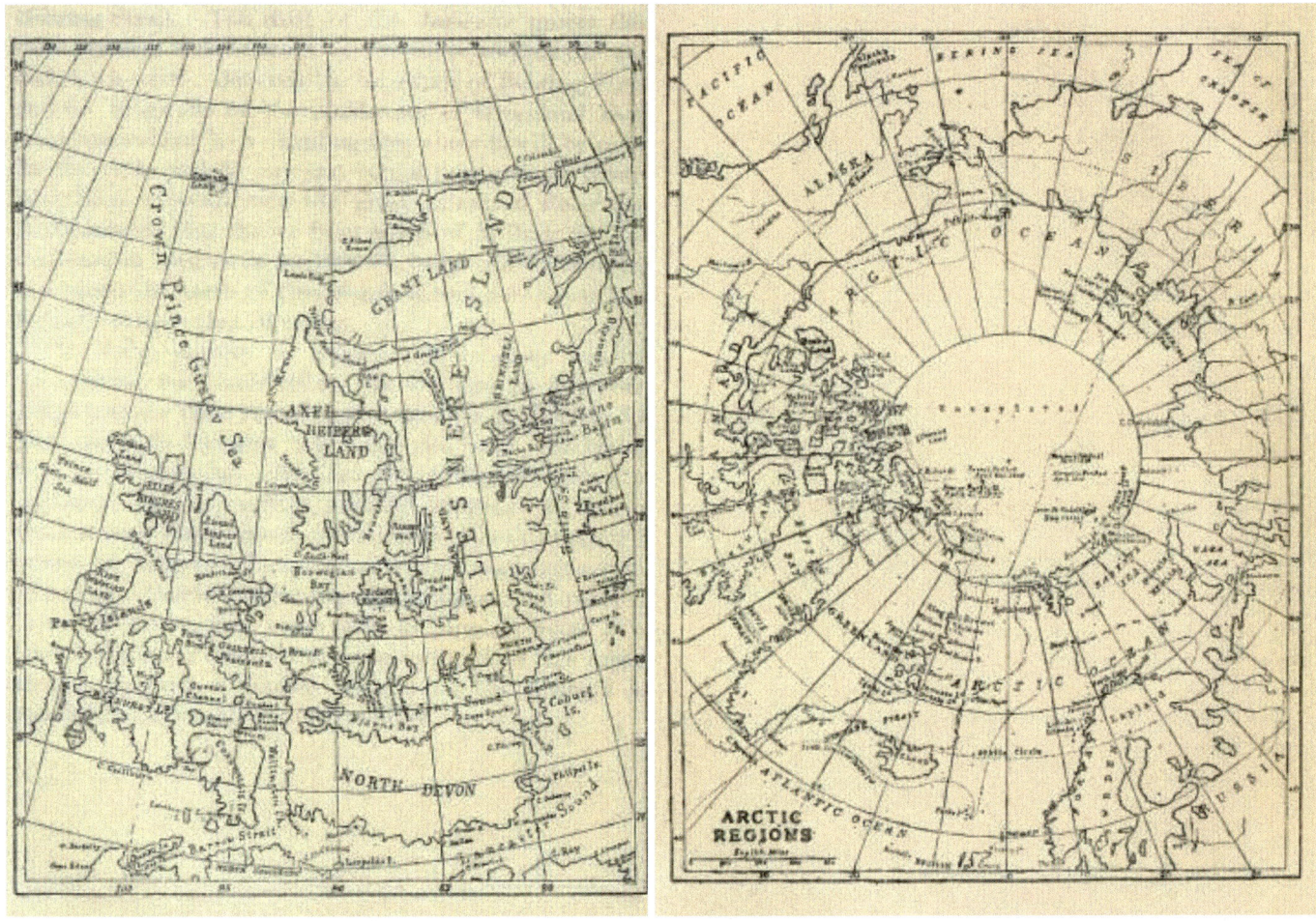

Chart of Sverdrup's Discoveries.   The Arctic regions.

Maps from *The Siege and Conquest of the North Pole,* by George Bryce, Projec Gutenberg edition (orginally published 1910 Gibbings & Co. Ltd., London)

Chapter 9: History and Mystery

No book on Axel Heiberg Island, or the Queen Elizabeth Islands in general, would be complete without a mention of some of the long-standing human and natural mysteries surrounding the region. Because of the difficulty of exploration and the extreme environmental conditions, answers to the mysteries come slowly, if at all.

The Missing Explorers

Hans Krüger, a German explorer fresh from an expedition in Greenland that had left him sore and weak from an attack of trichinonsis, set off to navigate the Canadian Arctic in 1930. His stated goals were geological, but many felt he had more ambitious aims. By all accounts he was underprepared and reluctant to listen to advice, setting off with overburdened sledges (komitik) that the dogs couldn't pull, requiring help from the human explorers to even get moving. Two weeks into the journey, the support sledges returned from Depot Point on Axel Heiberg back to Ellesmere; only two others accompanied him as they headed for the northern tip of Axel Heiberg.

By the spring of 1931 there was no sign of the party, and the Royal Canadian Mounted Police launched a series of hazardous search patrols. They were not able to reach Axel Heiberg that year, so efforts resumed the following spring. Corporal Stallworthy led an arduous dogsled circumnavigation of Axel Heiberg, the longest patrol ever conducted by the Royal North West Mounted Police and a journey that almost cost him his own life as his team ran critically low on food. He almost certainly would have starved to death had he not encountered a cache of food left by another party.

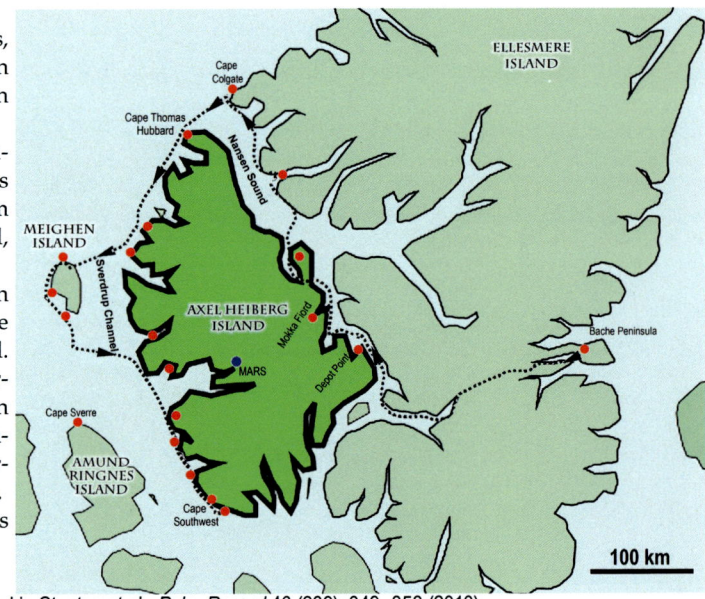

Arctic explorers would leave notes in cairns along their routes, and Stallworthy's patrol found one at Cape Hubbard, the northern end of Axel Heiberg. It stated Krüger's intentions to visit Meighen Island to the southeast.

The logical conclusion was that the party had died of starvation sometime in 1930 or 1931, although a satisfactory answer was never found. Krüger's mother wrote to the RCMP imploring them to continue the search for her son, and his fiancée never married, tragically taking her own life in 1946.

Two more notes were found in 1950, including one on Meighan Island that confirmed that the party had made it that far. That note indicated the next destination would be Amund Ringnes Island. No traces were ever found there, leading to theories that the overburdened sledges had broken through the ice between Meighan and Amund Ringnes, plunging the team to their deaths or condemning them to starvation. Others hypothesized that the explorers might have died of carbon monoxide poisoning in their tents.

Others, including Stallworthy, were skeptical of these theories and predicted that a "last camp" would someday be found.

The explorers' route as reconstructed in 2010. Image redrawn from data published in Stenton et al., *Polar Record* 46 (239): 349–358 (2010).

More clues almost seventy years later

In 1999, 3 geologists working near Cape Southwest came upon some old, partially buried materials that were clearly part of a camp or cache, including a large food canister. They took only a transit and compass for aid in identification, and reported their findings to the Canada Science and Technology Museum and the government of Nunavut. An archeological team visited the site for careful excavation in summer 2004. They found that the "food canister" contained only geological samples. Clothing and eating utensils suggested one man's items had been left behind, so it is thought that one member of the expedition had already died by this point. There were no tents, so this was not a camp but a cache, presumably one which the explorers hoped to retrieve at a later date.

The archaelogists continued to work with the retrieved materials, painstakingly opening slips of paper to preserve as many of the notes and labels as possible. Based upon the writings and the previous notes, an updated trace of the explorers' route was constructed and published in 2009.

Kruger's last camp remains undiscovered. It is possible that the explorers perished nearby while searching for food, but this is purely speculation. Visitors and researchers to the area are all advised to be on the lookout for any objects or materials that might give further hints into the team's final fate.

Document left by Krüger left at Cape Thomas Hubbard and discovered by Corporal Stallworthy in 1932.

Hans Krüger (on the right) with officers and men from *Beothic*, 1929.

These two images were reproduced with permission from Cambridge University Press from the article "The career and disappearance of Hans K.E. Krüger, Arctic geologist, 1886–1930," William Barr, *Polar Record*, 277-304 (1993). (Barr also wrote a biography of Stallworthy: see Bibliography).

Air travel is necessary in the Arctic, but hazardous due to the extreme terrain and weather and the unreliability of magnetic compasses so close to the Magnetic North Pole. The Twin Otter aircraft is a workhorse because of its ability to land on the tundra and to operate safely if one engine fails.

The sight of remains of aircraft accidents, such as this one in Resolute Bay, are not uncommon.

The Mummified Forest

Explorers on Axel Heiberg have known about the remains of large trees found in an area north of MARS since at least the 1960s. But it wasn't until xxx that the first scientific paper appeared describing the remains of a subtropical forest. Palms, cycads, cypress and redwoods lived and died on the island during the Eocene epoch (56 to 34 million years ago), leaving wood in a remarkable state of preservation: not petrified, but preserved to the point where it can still burn. It was preserved this way by being covered with silt during a flood, never becoming mineralized.

How could these plants grow in the Arctic? One's first guess might be that Axel Heiberg has moved due to plate tectonics. But tectonic records show that the paleolatitude of the island was the same as it is now, spanning the 80th parallel. The preserved trees also show growth patterns consistent with 6 months of darkness in the winter.

So during the Eocene, the night and day patterns at 80 degrees North latitude were the same as they are today. Neither the position of the island nor the obliquity of the Earth's orbit have changed. Nonetheless, the climate of the Arctic was balmy then, with an average yearly temperature of 18 degrees C (64 F, comparable to Baltimore, Maryland) compared with -10 degrees C (14 F) today.

Since the discovery of the forest, fossils of many subtropical animals have also been discovered, including turtles, alligators, large birds, and mammals resembling elephants.

This climate was like nothing we know on Earth today: six months of darkness, yet no ice, probably not even freezing temperatures. The "Eocene greenhouse" may have been related to carbon dioxide release from thawing permafrost, and warm temperatures preserved over winter by high humidity levels as seen in today's temperate rainforests.

Visitors are discouraged from visiting the forest and its location is kept as discreet as possible, since there are fears that casual taking of souvenirs will deplete this incredible resource. The author has never had a chance to see it, and so there are no pictures here. This remains something for the Bucket List.

Graffiti found on Axel Heiberg

This plaque found in Resolute commemorates the expedition of Hyoichi Kohno, who vanished in 2001 during a ski-and-walk adventure that was to take him over 9300 miles from the North Pole to his home town in Japan. When he failed to rendezvous by radio, rescuers went to search for him and found his sled near open water, and a few days later his body. Kohno had lived a life of adventure between the ages of 19 and his death at 43, trekking across the United States, across the Sahara, and up Mt. Aconcagua. The beginning of his North Pole journey marked him as the first Japanese to reach the North Pole on foot.

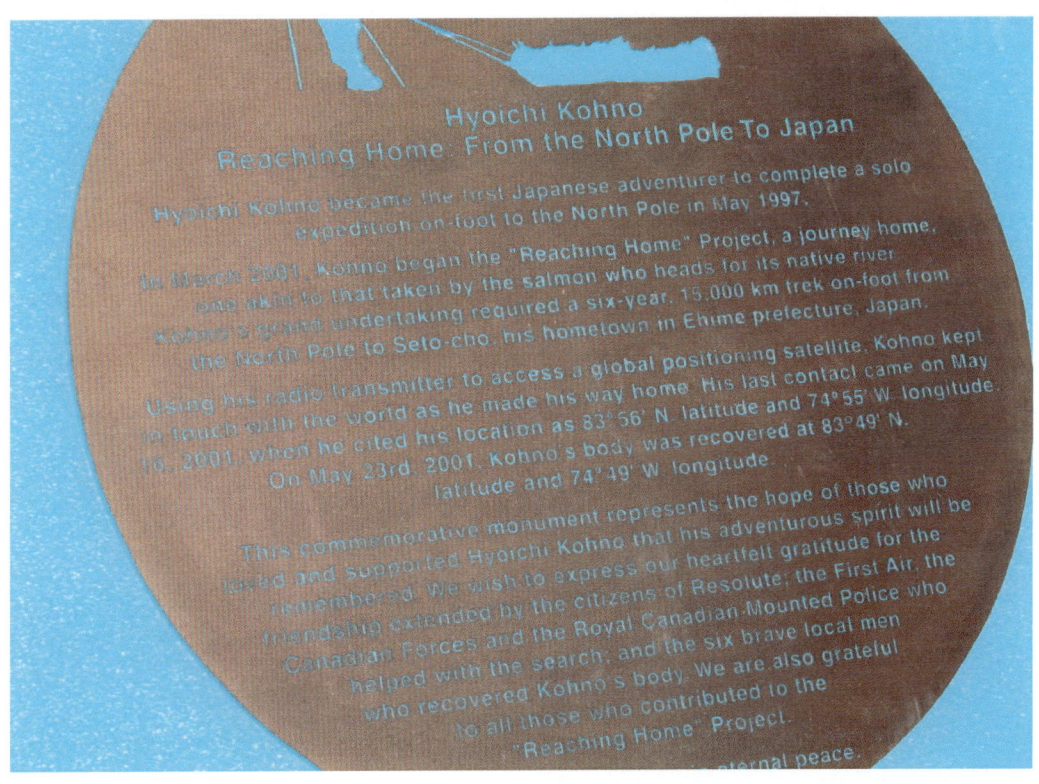

Hyoichi Kohno
Reaching Home: From the North Pole To Japan

Hyoichi Kohno became the first Japanese adventurer to complete a solo expedition on-foot to the North Pole in May 1997.

In March 2001, Kohno began the "Reaching Home" Project, a journey home, one akin to that taken by the salmon who heads for its native river. Kohno's grand undertaking required a six-year, 15,000 km trek on-foot from the North Pole to Seto-cho, his hometown in Ehime prefecture, Japan.

Using his radio transmitter to access a global positioning satellite, Kohno kept in touch with the world as he made his way home. His last contact came on May 16, 2001, when he cited his location as 83° 56' N. latitude and 74°55' W. longitude. On May 23rd, 2001, Kohno's body was recovered at 83°49' N. latitude and 74°49' W. longitude.

This commemorative monument represents the hope of those who loved and supported Hyoichi Kohno that his adventurous spirit will be remembered. We wish to express our heartfelt gratitude for the friendship extended by the citizens of Resolute, the First Air, the Canadian Forces and the Royal Canadian Mounted Police who helped with the search, and the six brave local men who recovered Kohno's body. We are also grateful to all those who contributed to the "Reaching Home" Project.

... eternal peace.

Parting Words and Acknowledgements

I was profoundly affected by my time in the Arctic, and feel very lucky to have had the privilege to visit. Those who love the Arctic often return again and again each season, drawn by the 24-hour daylight, the harsh contrast between summer's richness and winter's unfathomable cold, and the remoteness of some of Earth's last wilderness areas.

None of this would have been possible without financial support from CARN and from the Polar Continental Shelf Program. I also wish to provide a sincere thanks to all of the support staff in Resolute Bay who went above and beyond to help us get to and from Axel Heiberg as well as to perform sample processing in the laboratories there. A special shout-out to the kitchen crew, who not only made great food, but who provided packets of gelatin so we could make coated slides for an experiment.

To all of the people who helped me and my team at MARS, many of whom appear in this book, thank you for all of your support and guidance and for making MARS the haven that it is. On my first trip, I thought we'd be bringing all of our own food, not realizing the extent of the fresh and preserved stocks in the MARS kitchen or the cooking skills of its residents. Wayne Pollard and Miles Ecclestone did so much to keep MARS running for everyone.

To Lyle Whyte for pointing me to CARN in the first place, for years of collaboration, and for his contributions to the research section and to all of the photos of MARS in May. Lyle brings a team to Axel Heiberg at least once a year. His team members Nancy Perrault, Tom Neiderberger, and Melissa Battler all appear here, and I am grateful for their permission to use their images.

Several of my own trainees appear here. Eve-Marie Dumas completed her M.Eng. degree at McGill based upon the work performed here with the holographic microscope. Philippe Giguère did his PhD at McGill focused on the robot shown here; he is now a professor at Université Laval. Jeremy Rogers traveled with me the first year to work on fluorescence microscopy. He took many of the photos here, including the beautiful close-ups of the ruddy turnstones and the action shots of the planes coming in to land. He is now a professor at the University of Wisconsin Madison.

Bibliography
Books

Barr, W. *Red Serge and Polar Bear Pants: The Biography of Harry Stallworthy, RCMP.* (University of Alberta Press, Edmonton; 2004).

Bocking, S. & Heidt, D. *Cold science : environmental knowledge in the North American Arctic during the Cold War.* (Routledge, London; New York; 2019).

Dick, L. *Muskox land : Ellesmere Island in the age of contact.* (University of Calgary Press, Calgary; 2001).

Jenness, D., Jenness, S.E. & Canadian Museum of Civilization. *Arctic odyssey : the diary of Diamond Jenness, ethnologist with the Canadian Arctic Expedition in Northern Alaska and Canada, 1913-1916.* (Canadian Museum of Civilization, Hull, Quebec; 1991).

Jenness, S.E. & Canadian Museum of Civilization. *Stefansson, Dr. Anderson and the Canadian Arctic Expedition, 1913-1918 : a story of exploration, science and sovereignty.* (Canadian Museum of Civilization, Gatineau, Québec; 2011).

Lopez, B.H. *Arctic dreams : imagination and desire in a northern landscape*, Edn. 1st Vintage Books. (Vintage Books, New York; 2001).

Osborne, S.L. *In the shadow of the pole : an early history of Arctic expeditions, 1871-1912.* (Dundurn Press, Toronto; 2013).

Pigott, P. *From far and wide : a complete history of Canada's Arctic sovereignty.* (Dundurn, Toronto; 2011).

Taylor, A. *Geographical Discovery and Exploration in the Queen Elizabeth Islands.* (E. Cloutier, 1955).

Weale, J. & Hassell, H. *Patterns to infinity : a Canadian artist's voyage to the Arctic.* (Fednav Ltd., Montreal; 1994).

Kobalenko, J. *The horizontal Everest : extreme journeys on Ellesmere Island.* (Soho, New York, NY; 2002).

Journal Articles: MARS Science

Adams, W.P., Cogley, J.G., Ecclestone, M.A. & Demuth, M.N. A small glacier as an index of regional mass balance: Baby Glacier, Axel Heiberg Island, 1959-1992. Geografiska Annaler Series a-Physical Geography 80A, 37-50 (1998).

Aitken, A.E. & Gilbert, R. Marine mollusca from expedition Fiord, western axel Heiberg Island, northwest territories, Canada. Arctic 49, 29-43 (1996).

Andersen, D.T., Pollard, W.H., McKay, C.P. & Heldmann, J. Cold springs in permafrost on Earth and Mars. Journal of Geophysical Research-Planets 107 (2002).

Battler, M.M., Osinski, G.R. & Banerjee, N.R. Mineralogy of saline perennial cold springs on Axel Heiberg Island, Nunavut, Canada and implications for spring deposits on Mars. Icarus 224, 364-381 (2013).

Chauhan, A. et al. Metagenomes from Thawing Low-Soil-Organic-Carbon Mineral Cryosols and Permafrost of the Canadian High Arctic. Microbiology Resource Announcements 2 (2014).

Cogley, J.G., Adams, W.P. & Ecclestone, M.A. Half a Century of Measurements of Glaciers on Axel Heiberg Island, Nunavut, Canada. Arctic 64, 371-375 (2011).

Cogley, J.G., Adams, W.P., Ecclestone, M.A., JungRothenhausler, F. & Ommanney, C.S.L. Mass balance of White Glacier, Axel Heiberg island, NWT, Canada, 1960-91. Journal of Glaciology 42, 548-563 (1996).

Colangelo-Lillis, J., Wing, B.A. & Whyte, L.G. Low viral predation pressure in cold hypersaline Arctic sediments and limits on lytic replication. Environmental Microbiology Reports 8, 250-260 (2016).

Doran, P.T. Sedimentology of Color Lake, a nonglacial high Arctic Lake, Axel-Heiberg Island, NWT, Canada. Arctic and Alpine Research 25, 353-367 (1993).

Fox-Powell, M.G. et al. Natural Analogue Constraints on Europa's Non-ice Surface Material. Geophysical Research Letters 46, 5759-5767 (2019).

Goordial, J. et al. In Situ Field Sequencing and Life Detection in Remote (79 degrees 26 ' N) Canadian High Arctic Permafrost Ice Wedge Microbial Communities. Frontiers in Microbiology 8 (2017).

Haltigin, T.W., Pollard, W.H., Dutilleul, P. & Osinski, G.R. Geometric Evolution of Polygonal Terrain Networks in the Canadian High Arctic: Evidence of Increasing Regularity over Time. Permafrost and Periglacial Processes 23, 178-186 (2012).

Heldmann, J.L., Pollard, W.H., McKay, C.P., Andersen, D.T. & Toon, O.B. Annual development cycle of an icing deposit and associated perennial spring activity on Axel Heiberg Island, Canadian High Arctic. Arctic Antarctic and Alpine Research 37, 127-135 (2005).

Heyer, E. Climate and energy-balance on Arctic tundra-Axel-Heiberg Island, Canadian Arctic Archipelago-Dhmura, A. Petermanns Geographische Mitteilungen 129, 307-307 (1985).

Jericho, S.K. et al. In-line digital holographic microscopy for terrestrial and exobiological research. Planetary and Space Science 58, 701-705 (2010).

Johannesson, K.H. & Lyons, W.B. Rare-earth element geochemistric of Color Lake, an acidic fresh-water lake on Axel-Heiberg-Island, Northwest-Territories, Canada. Chemical Geology 119, 209-223 (1995).

Jones, M.K.W., Pollard, W.H. & Jones, B.M. Rapid initialization of retrogressive thaw slumps in the Canadian high Arctic and their response to climate and terrain factors. Environmental Research Letters 14 (2019).

Macey, M.C. et al. The identification of sulfide oxidation as a potential metabolism driving primary production on late Noachian Mars. Scientific Reports 10 (2020).

Marcolefas, E. et al. Culture-Dependent Bioprospecting of Bacterial Isolates From the Canadian High Arctic Displaying Antibacterial Activity. Frontiers in Microbiology 10 (2019).

Medrzycka, D., Copland, L., Van Wychen, W. & Burgess, D. Seven decades of uninterrupted advance of Good Friday Glacier, Axel Heiberg Island, Arctic Canada. Journal of Glaciology 65, 440-452 (2019).

Michelutti, N., Douglas, M.S.V., Muir, D.C.G., Wang, X.W. & Smol, J.P. Limnological characteristics of 38 lakes and ponds on Axel Heiberg Island, High Arctic Canada. International Review of Hydrobiology 87, 385-399 (2002).

Moisan, Y. & Pollard, W. Formative process of the Thompson Glacier push structure, Axel Heiberg Island (Northwest Territories). Canadian Geographer-Geographe Canadien 39, 58-68 (1995).

Mueller, D.R. & Pollard, W.H. Gradient analysis of cryoconite ecosystems from two polar glaciers. Polar Biology 27, 66-74 (2004).

Muller, F. Was Good Friday Glacier on Axel Heiberg Island Surging. Canadian Journal of Earth Sciences 6, 891-+ (1969).

Nadeau, J.L. et al. Fluorescence Microscopy as a Tool for In Situ Life Detection. Astrobiology 8, 859-875 (2008).

Niederberger, T.D. et al. Novel sulfur-oxidizing streamers thriving in perennial cold saline springs of the Canadian high Arctic. Environmental Microbiology 11, 616-629 (2009).

Ohmura, A. Evaportation from the surface of the Arctic tundra on Axel-Heiberg Island. Water Resources Research 18, 291-300 (1982).

Omelon, C.R., Pollard, W.H. & Andersen, D.T. A geochemical evaluation of perennial spring activity and associated mineral precipitates at Expedition Fjord, Axel Heiberg Island, Canadian High Arctic. Applied Geochemistry 21, 1-15 (2006).

Perreault, N.N., Andersen, D.T., Pollard, W.H., Greer, C.W. & Whyte, L.G. Characterization of the prokaryotic diversity in cold saline perennial springs of the Canadian high Arctic. Applied and Environmental Microbiology 73, 1532-1543 (2007).

Perreault, N.N. et al. Heterotrophic and Autotrophic Microbial Populations in Cold Perennial Springs of the High Arctic. Applied and Environmental Microbiology 74, 6898-6907 (2008).

Pollard, W. et al. Overview of analogue science activities at the McGill Arctic Research Station, Axel Heiberg Island, Canadian High Arctic. Planetary and Space Science 57, 646-659 (2009).

Pollard, W., Omelon, C., Andersen, D. & McKay, C. Perennial spring occurrence in the Expedition Fiord area of western Axel Heiberg Island, Canadian High Arctic. Canadian Journal of Earth Sciences 36, 105-120 (1999).

Pollard, W.H. Icing processes associated with high arctic perennial springs, Axel Heiberg Island, Nunavut, Canada. Permafrost and Periglacial Processes 16, 51-68 (2005).

Rusley, C. et al. Metagenome-Assembled Genome of USC alpha AHI, a Potential High-Affinity Methanotroph from Axel Heiberg Island, Canadian High Arctic. Microbiology Resource Announcements 8 (2019).

Samson, C. et al. Combined electromagnetic geophysical mapping at Arctic perennial saline springs: Possible applications for the detection of water in the shallow subsurface of Mars. Advances in Space Research 59, 2325-2334 (2017).

Sapers, H.M. et al. Biological Characterization of Microenvironments in a Hypersaline Cold Spring Mars Analog. Frontiers in Microbiology 8 (2017).

Schuerger, A.C. & Nicholson, W.L. Twenty Species of Hypobarophilic Bacteria Recovered from Diverse Soils Exhibit Growth under Simulated Martian Conditions at 0.7 kPa. Astrobiology 16, 964-976 (2016).

Simoneit, B.R.T., Otto, A., Kusumoto, N. & Basinger, J.F. Biomarker compositions of Glyptostrobus and Metasequoia (Cupressaceae) fossils from the Eocene Buchanan Lake Formation, Axel Heiberg Island, Nunavut, Canada reflect diagenesis from terpenoids of their related extant species. Review of Palaeobotany and Palynology 235, 81-93 (2016).

Thomson, L.I. & Copland, L. Changing contribution of peak velocity events to annual velocities following a multi-decadal slowdown at White Glacier. Annals of Glaciology 58, 145-154 (2017).

Thomson, L.I. & Copland, L. Multi-decadal reduction in glacier velocities and mechanisms driving deceleration at polythermal White Glacier, Arctic Canada. Journal of Glaciology 63, 450-463 (2017).

Thomson, L.I., Osinski, G.R. & Ommanney, C.S.L. Glacier change on Axel Heiberg Island, Nunavut, Canada. Journal of Glaciology 57, 1079-1086 (2011).

Thomson, L.I., Zemp, M., Copland, L., Cogley, J.G. & Ecclestone, M.A. Comparison of geodetic and glaciological mass budgets for White Glacier, Axel Heiberg Island, Canada. Journal of Glaciology 63, 55-66 (2017).

Trivedi, C.B., Lau, G.E., Grasby, S.E., Templeton, A.S. & Spear, J.R. Low-Temperature Sulfidic-Ice Microbial Communities, Borup Fiord Pass, Canadian High Arctic. Frontiers in Microbiology 9 (2018).

Van Wychen, W. et al. Characterizing interannual variability of glacier dynamics and dynamic discharge (1999-2015) for the ice masses of Ellesmere and Axel Heiberg Islands, Nunavut, Canada. Journal of Geophysical Research-Earth Surface 121, 39-63 (2016).

Ward, M.K. & Pollard, W.H. A hydrohalite spring deposit in the Canadian high Arctic: A potential Mars analogue. Earth and Planetary Science Letters 504, 126-138 (2018).

Wilhelm, R.C., Niederberger, T.D., Greer, C. & Whyte, L.G. Microbial diversity of active layer and permafrost in an acidic wetland from the Canadian High Arctic. Canadian Journal of Microbiology 57, 303-315 (2011).

Zentilli, M., Omelon, C.R., Hanley, J. & LeFort, D. Paleo-Hydrothermal Predecessor to Perennial Spring Activity in Thick Permafrost in the Canadian High Arctic, and Its Relation to Deep Salt Structures: Expedition Fiord, Axel Heiberg Island, Nunavut. Geofluids (2019).

Journal Articles: Natural History of Axel Heiberg Island

Bigras, C., Bilz, M., Grattan, D.W. & Gruchy, C. Erosion of the Geodetic Hills Fossil Forest, Axel Heiberg Island, Northwest Territories. Arctic 48, 342-353 (1995).

Blanchette, R.A., Cease, K.R., Abad, A.R., Burnes, T.A. & Obst, J.R. Ultrastructural characterization of wood from tertiary fossil forests in the Canadian High Arctic. Canadian Journal of Botany-Revue Canadienne De Botanique 69, 560-568 (1991).

Bono, R.K., Clarke, J., Tarduno, J.A. & Brinkman, D. A Large Ornithurine Bird (Tingmiatornis arctica) from the Turonian High Arctic: Climatic and Evolutionary Implications. Scientific Reports 6 (2016).

Dolezych, M. & Reinhardt, L. First evidence for the conifer Pinus, as Pinuxylon selmeierianum sp. nov., during the Paleogene on Wootton Peninsula, northern Ellesmere Island, Nunavut, Canada. Canadian Journal of Earth Sciences 57, 25-39 (2020).

Eberle, J.J. & Greenwood, D.R. Life at the top of the greenhouse Eocene world-A review of the Eocene flora and vertebrate fauna from Canada's High Arctic. Geological Society of America Bulletin 124, 3-23 (2012).

Eberle, J.J. & Storer, J.E. Northernmost record of brontotheres, Axel Heiberg Island, Canada - Implications for age of the Buchanan Lake Formation and brontothere paleobiology. Journal of Paleontology 73, 979-983 (1999).

Friedman, M., Tarduno, J.A. & Brinkman, D.B. Fossil fishes from the high Canadian Arctic: further palaeobiological evidence for extreme climatic warmth during the Late Cretaceous (Turonian-Coniacian). Cretaceous Research 24, 615-632 (2003).

Gulbranson, E.L. et al. Leaf habit of Late Permian Glossopteris trees from high-palaeolatitude forests. Journal of the Geological Society 171, 493-507 (2014).

Jahren, A.H. The Arctic Forest of the Middle Eocene. Annual Review of Earth and Planetary Sciences 35, 509-540 (2007).

Lepage, B.A. A new species of Tsuga (Pinaceae) from the middle Eocene of Axel Heiberg Island, Canada, and an assessment of the evolution and biogeographical history of the genus. Botanical Journal of the Linnean Society 141, 257-296 (2003).

LePage, B.A. A new species of Thuja (Cupressaceae) from the Late Cretaceous of Alaska: Implications of being evergreen in a polar environment. American Journal of Botany 90, 167-174 (2003).

Leslie, A.B. & Pfefferkorn, H.W. Fossil floras from the Emma Fiord Formation (Visean, Mississippian) of the Canadian Arctic Archipelago and their paleoenvironmental context. Review of Palaeobotany and Palynology 159, 195-203 (2010).

Maxbauer, D.P., Royer, D.L. & LePage, B.A. High Arctic forests during the middle Eocene supported by moderate levels of atmospheric CO_2. Geology 42, 1027-1030 (2014).

McIver, E.E. & Basinger, J.F. Early Tertiary floral evolution in the Canadian high arctic. Annals of the Missouri Botanical Garden 86, 523-545 (1999).

Mustoe, G.E. Non-Mineralized Fossil Wood. Geosciences 8 (2018).

Schoenhut, K. Environmental implications of the preservation of chloroplast ultrastructure in Eocene Metasequoia leaves. Paleobiology 31, 424-433 (2005).

Schoenhut, K., Vann, D.R. & LePage, B.A. Cytological and ultrastructural preservation in eocene Metasequoia leaves from the Canadian high arctic. American Journal of Botany 91, 816-824 (2004).

Staccioli, G., McMillan, N.J., Meli, A. & Bartolini, G. Chemical characterisation of a 45 million year bark from geodetic hills fossil forest, Axel Heiberg Island, Canada. Wood Science and Technology 36, 419-427 (2002).

Tarduno, J.A. et al. Evidence for extreme climatic warmth from Late Cretaceous Arctic vertebrates. Science 282, 2241-2244 (1998).

Vandermark, D., Tarduno, J.A. & Brinkman, D.B. Late Cretaceous plesiosaur teeth from Axel Heiberg Island, Nunavut, Canada. Arctic 59, 79-82 (2006).

Vandermark, D., Tarduno, J.A. & Brinkman, D.B. A fossil champsosaur population from the high Arctic: Implications for Late Cretaceous paleotemperatures. Palaeogeography Palaeoclimatology Palaeoecology 248, 49-59 (2007).

Vavrek, M.J., Hills, L.V. & Currie, P.J. A Hadrosaurid (Dinosauria: Ornithischia) from the Late Cretaceous (Campanian) Kanguk Formation of Axel Heiberg Island, Nunavut, Canada, and Its Ecological and Geographical Implications. Arctic 67, 1-9 (2014).

Vavrek, M.J., Larsson, H.C.E. & Rybczynski, N. A Late Triassic flora from east-central Axel Heiberg Island, Nunavut Canada. Canadian Journal of Earth Sciences 44, 1653-1659 (2007).

West, C.K., Greenwood, D.R. & Basinger, J.F. Was the Arctic Eocene 'rainforest' monsoonal? Estimates of seasonal precipitation from early Eocene megafloras from Ellesmere Island, Nunavut. Earth and Planetary Science Letters 427, 18-30 (2015).

West, C.K., Greenwood, D.R. & Basinger, J.F. The late Paleocene to early Eocene Arctic megaflora of Ellesmere and Axel Heiberg islands, Nunavut, Canada. Palaeontographica Abteilung B-Palaeophytologie Palaeobotany-Palaeophytology 300, 47-163 (2019).

Parker, G.R. Morphology, reproduction, diet, and behavior of Arctic hare (Lepus arcticus monstrabilis) on Axel Heiberg Island, Northwest-Territories. Canadian Field-Naturalist 91, 8-18 (1977).

Journal and News Articles: History and Exploration

Grove, J.M. Preliminary-report of the Jacobsen-McGill Arctic research expedition to Axel Heiberg Island-Muller, F. Geographical Journal 128, 230-231 (1962).

Park, R.W. & Stenton, D.R. A Hans Kruger Arctic expedition cache on Axel Heiberg Island, Nunavut. Arctic 60, 1-6 (2007).

Stenton, D.R., Park, R.W. & Grant, T. Retracing the route of Hans K. E. Kruger's 1930 German Arctic expedition. Polar Record 46, 349-358 (2010).

Brooks, R.C., et al. Krüger's final camp in Arctic Canada? Arctic 57(2):225 – 229 (2004).

D. L. Brown "An Unfinished Trek Through the Arctic: In Canada, Mourners Remember Japanese Man Who Died Doing What He Loved." Washington Post Foreign Service Sunday, June 3, 2001; Page A20

Walker, N. "Team follows Norwegian explorer's Arctic expedition: Retracing the treks of famed Norwegian Arctic explorer Otto Sverdrup." Canadian Geographic, April 1, 2014.

Struzik, E. "Th e Arctic is a book of untold stories." Edmonton Journal, December 12, 2010.